A Moscow Math Circle

Week-by-week Problem Sets

A Moscow Math Circle

Week-by-week Problem Sets

Sergey Dorichenko

Translated by Tatiana Shubin

Mathematical Sciences Research Institute
Berkeley, California

American Mathematical Society
Providence, Rhode Island

Advisory Board for the MSRI/Mathematical Circles Library

David Auckly
Zuming Feng
Tony Gardiner
Kiran Kedlaya
Nikolaj N. Konstantinov
Silvio Levy
Walter Mientka
Bjorn Poonen

Alexander Shen
Tatiana Shubin (Chair)
Zvezdelina Stankova
Ravi Vakil
Ivan Yashchenko
Paul Zeitz
Joshua Zucker

Series Editor: Silvio Levy, Mathematical Sciences Research Institute

This volume is published with the generous support of the John Templeton Foundation.

Original work © 2011 Sergey Dorichenko
Translated by Tatiana Shubin.
Translation edited by David Scott and Silvio Levy.

The assortment of cartoons appearing at various places in this book were previously published in *Kvant* and are used with the permission of Eugene Nazarov.

2010 *Mathematics Subject Classification*. Primary 00A08; Secondary 00A07.

For additional information and updates on this book, visit
www.ams.org/bookpages/mcl-8

Library of Congress Cataloging-in-Publication Data
Dorichenko, S. A. (Sergei Aleksandrovich), 1973–
  A Moscow math circle : week-by-week problem sets / Sergey Dorichenko ; translated by Tatiana Shubin ; translation edited by David Scott and Silvio Levy.
    p. cm. — (MSRI mathematical circles library ; 8)
  ISBN 978-0-8218-6874-4 (alk. paper)
  1. Mathematics—Problems, exercises, etc. I. Title.
QA43.D57 2011
510—dc23
                                                                              2011036422

Copying and reprinting. Individual readers of this publication, and nonprofit libraries acting for them, are permitted to make fair use of the material, such as to copy a chapter for use in teaching or research. Permission is granted to quote brief passages from this publication in reviews, provided the customary acknowledgment of the source is given.

Republication, systematic copying, or multiple reproduction of any material in this publication is permitted only under license from the American Mathematical Society. Requests for such permission should be addressed to the Acquisitions Department, American Mathematical Society, 201 Charles Street, Providence, Rhode Island 02904-2294 USA. Requests can also be made by e-mail to reprint-permission@ams.org.

© 2012 by the Mathematical Sciences Research Institute. All rights reserved.
Printed in the United States of America.

∞ The paper used in this book is acid-free and falls within the guidelines
established to ensure permanence and durability.
Visit the AMS home page at http://www.ams.org/
Visit the MSRI home page at http://www.msri.org/

Contents

Foreword	ix
Preface	xi
Acknowledgments	xxiii
Problem Set 0	1
Problem Set 1	3
Problem Set 2	5
Problem Set 3	7
Problem Set 4	9
Problem Set 5	11
Problem Set 6	13
Problem Set 7	15
Problem Set 8	17
Problem Set 9	19
Problem Set 10	21
Problem Set 11	23
Problem Set 12	25
Problem Set 13	27
Winter Competition	29
Problem Set 14	32
Problem Set 15	34

Problem Set 16	36
Problem Set 17	38
Problem Set 18	40
Problem Set 19	42
Problem Set 20	44
Problem Set 21	46
Problem Set 22	48
Problem Set 23	50
Problem Set 24	52
Problem Set 25	54
Problem Set 26	56
Problem Set 27	58
Problem Set 28	60
Solutions to Problem Set 0	63
Solutions to Problem Set 1	66
Solutions to Problem Set 2	69
Solutions to Problem Set 3	73
Solutions to Problem Set 4	78
Solutions to Problem Set 5	83
Solutions to Problem Set 6	87
Solutions to Problem Set 7	90
Solutions to Problem Set 8	95
Solutions to Problem Set 9	98
Solutions to Problem Set 10	102
Solutions to Problem Set 11	106
Solutions to Problem Set 12	114
Solutions to Problem Set 13	118
Solutions to the Winter Competition	121

Contents	vii
Solutions to Problem Set 14	130
Solutions to Problem Set 15	134
Solutions to Problem Set 16	140
Solutions to Problem Set 17	144
Solutions to Problem Set 18	147
Solutions to Problem Set 19	153
Solutions to Problem Set 20	159
Solutions to Problem Set 21	164
Solutions to Problem Set 22	168
Solutions to Problem Set 23	173
Solutions to Problem Set 24	176
Solutions to Problem Set 25	179
Solutions to Problem Set 26	182
Solutions to Problem Set 27	185
Solutions to Problem Set 28	188
Mathematical Maze	193
Two and Two Is More Than Four: A Story	205
Addendum: The San Jose Experience by Tatiana Shubin	215
Problem Set SJ1	219
Solutions to Problem Set SJ1	221
Problem Set SJ2	227
Solutions to Problem Set SJ2	229
Problem Set SJ3	235
Solutions to Problem Set SJ3	237

Foreword

Many people involved with math circles know and highly regard two remarkable books: *Mathematical Circles (Russian Experience)* and *A Decade of the Berkeley Math Circle: The American Experience* (Volume 1). Both books are organized by topic.

In contrast, Sergei Dorichenko's *A Moscow Math Circle: Week-by-Week Problem Sets* has a very different structure. As suggested by the title, it consists mostly of transcriptions of a year of math circle meetings for eighth-grade Moscow students. The author describes this circle in detail in his Preface; there is no need to repeat what is said there. Instead, I want to explain why this book has been selected to be among the first published in the MSRI Math Circles Library.

In running a math circle, choosing a suitable topic and a set of good problems is as crucial as it is difficult, even for experienced leaders. To have a coherent collection that can be used for an entire year is a dream coming true.

In this book, problems for each meeting are chosen with great care: at each set, there are several topics and techniques at play. Typically, one or more new topic or technique is introduced, while there are a few problems dealing with topics that have been introduced earlier. In the latter case, problems could call for a slight variation of the previously learned technique or for a different viewpoint on an old topic.

It can be very enjoyable to use these sets of problems, meeting after meeting, in the order in which they are presented. At the same time, each set is self-contained and can be used completely independently: a nice way to fill in an unexpected hole in a circle's schedule!

Additional problems, presented in the Winter Competition (page 29), the contest known as the Mathematical Maze (page 193), and an editorial addendum (page 215), enhance this substantial collection even further. And, unusually for a problem book, there is also a short story, "Two and Two Is More Than Four" (page 205), based on actual discussions held by the author's students.

It is hoped that the story will serve a couple of purposes. First, it's good for kids: it might catch someone who's not yet especially keen on problem solving. Such students wouldn't start by dealing with the problem sets, but they might find reading the story amusing and thus be lured into the rest of the book. Second, the story is good for adults: the difference between tutoring a kid — even a very smart one — and leading a circle lies precisely in that kids get to interact with one another in the circle, and it is immensely rewarding to watch them do so. This story captures some of this free play of ideas and emerging discoveries. Thus it might whet someone's appetite and even prompt them to become a circle leader.

Finally, the author's Preface is brimming with interesting pedagogical and organizational ideas. He describes two Moscow circles built on a rich century-long tradition, and he explains how they operate and what draws hundreds of young students to the weekly meetings. Implementing some of these ideas might enhance and enliven many U.S. circles for students of various ages.

Tatiana Shubin
Chair, MSRI/MCL Advisory Board
San Jose, July 2011

Preface

This book presents materials that were used during the course of a year in one of the math circles for eighth graders organized by members of the mathematics faculty at Moscow State University. At different times and with different groups of students these materials have also been used in a math circle run at Moscow School Number 57, a mathematics "magnet" school. Some of the students had attended the circle the previous year, but since most were novices, the sessions were generally aimed at beginners. This does not mean that the sessions were easy, for even very simple problems can frustrate a beginner. Some minimal knowledge from school was expected, but much school material needed reviewing. For more advanced and experienced students, there were always extra problems available.

Students and their parents learn about math circles by word of mouth, from the Internet, or at math Olympiads. For instance, at the end of September in Moscow there is the Lomonosov Tournament, a multisubject competition for sixth through eleventh graders.[1] This is a remarkable event. It is held on a Sunday in several Moscow colleges simultaneously, and it has recently spread to other cities. The rules and problems are the same for all sites. A student comes to one of the hosting sites and finds simultaneous contests in mathematics, math games, physics, biology, linguistics, history, astronomy, literature, and chemistry. Competing in any of these contests takes about an hour and can commence at virtually any moment during the Olympiad; the whole event takes about five hours. This gives students a chance, by moving from one room to another, to participate in several contests of their choice. Those who perform well in a given contest receive a winner certificate for this contest, and those who succeed in several contests get an all-around winner certificate.

But the certificates are not the most important goal; the real goal is to interest students in sciences and invite them to the circles. Students may dislike a given subject for all sorts of reasons—for example, if they happen to have a teacher who isn't very good—but when there are lots of contests happening simultaneously, their curiosity may force them to visit a contest, say, in math games, even if they are indifferent to mathematics. Their

---

[1]In the U.S., this would be seventh through twelfth graders.

attitude towards some subjects may change after the Lomonosov tournament since students see that every subject can be interesting. All the participants of the Tournament receive a brochure describing various Moscow circles.

Math circles have existed for more than a century at Moscow State University, and today they are combined under the name of "Little Mech-Mat" — an allusion to the nickname for the university's renowned mechanics and mathematics school. There are two divisions: one with evening sessions and another by correspondence. The evening division of the Little Mech-Mat comprises of circles for students in grades 6 through 11, although it isn't unusual for a fifth grader to attend a sixth graders circle. There is also a group for younger students. Anybody can participate in a math circle; there are no exams and all the circle sessions are free of charge.

The sessions are held in classrooms in the MechMat building on Saturday evenings when most of the rooms are vacant. In each classroom, there are 15 to 30 students and three to six instructors. Most instructors are university students and often they are former circle participants, but the head instructor is usually an experienced mathematician. The composition of the student group is fluid, as new students join all the time and some students drop out. Yet there is always a nucleus: a group of students who regularly attend the circle. The instructional staff does not change, except that some may not be able to attend every session. There is a winter break when students take exams and go on vacation.

The total number of participants in each grade from 6 through 8 is usually from 100 to 200 on any given evening. There are fewer students in the higher grades (9 through 11), since by this age many have become regular students of mathematics day schools (somewhat similar to "magnet schools" in the U.S.) and have demanding work loads there. Each grade has a head instructor who is responsible for designing the program for that particular grade. All instructors of a given grade may participate in developing class materials for that grade. Although individual sessions are held in several auditoriums, all sections of the same grade use the same handouts. One exception to this practice occurs when an instructor chooses to create his or her own individual course. In the higher grades such individual courses are more common, but the number of participants in each of them can be quite small.

The first class of the year is often given in written form because the organizers cannot anticipate how many students will show up and consequently don't know how many instructors will be necessary. Problems selected for the first session are not too difficult and do not require a lot of writing. Students are told that this is not a test: the main goal of the meeting is to find out what these students know, what they don't know, and what they need to learn. We tell them: the more you do not know, the better, because then there's more we can teach you. When they hear this, they break out smiling. We also tell the students that they can solve problems in any order, and that they should not be afraid of the prospect of solving just a few, and even if they can't solve any, it is still okay. If, on the other hand, some students solve all the problems quickly, they may be offered more challenging additional problems.

At the first meeting students also fill out a short questionnaire. At the next session each student finds his/her name, together with a specified classroom, on the list of circle participants. Students may change their classroom assignment if they wish.

In each subsequent session, a separate classroom is designated for that day's newcomers; later they are transferred to other classrooms. This way each regular classroom is periodically supplemented with newcomers. Session topics change quite often, so even students who join the circle in the middle of the year will see sets of problems with which they have a good chance of success.

Students vary considerably in strength and knowledge. There are several different approaches to dealing with this complication. Classrooms may be specifically designated for beginners and advanced students, with the level of the class materials accordingly different. This approach requires twice as much work, and quite often there are simply not enough resources to do this. Another possibility is to make only one version of the handout for all students, without sorting them into different levels. In this case every problem set must have very simple problems, more challenging ones, and also additional problems. This approach gives students an opportunity to move at their own pace, and it is quite manageable with several instructors in the room. Discussions at the board may also be done in a way that is useful for students at different levels.

Another wonderful approach is to put all new students into a big auditorium. There the students play fun games and solve easy problems that are carefully discussed. Those students who cope well with the easy problems are transferred to a regular math circle; the ones who struggle stay in the auditorium for newcomers. To be successful, this approach requires very good instructors for the newcomers.

The last sessions before the winter and summer breaks are very special. They may include a session on topological puzzles where students are tied with ropes and have to free themselves without cutting or untying the ropes. Another possibility is a competition which we call the Math Maze: it is described in detail starting on page 193, but here is the general idea.

Each student is given a map of the "maze" and the rules of the game. The map contains room numbers and their names, such as Mental Calculation, Games, Geometry, Logic, Combinatorics, or Puzzles. A student must visit all the rooms in any order and solve one problem in each room. The rooms are spread all over the building, so part of the fun is that some students like to race from one room to another. After a student has visited all the rooms on his map, he goes to the final auditorium where he can choose a book as a prize.

Winter break is quite long; sometimes we are able to shorten it by finding enthusiastic instructors willing to work with students even during their own final exams and vacations. But whatever the situation, before the break students are given a page with the Winter Competition problems, so that they do not forget the math circle. Students are expected to work on these problems over the break, write down their solutions and bring them to one of the first sessions after the break for grading. There are awards (usually math books). Everybody who turns in solutions gets one, but of course the higher the score, the greater the award.

The sessions at the Wednesday evening math circle at Moscow School Number 57, a day school with emphasis on math, are similar to the ones just described. About 100 students per grade participate; each room has 15 students and three to four instructors, including some older students from the school who are eager to help. The circle is for grades 7 and 8, though recently a new circle for sixth graders began. Their winter break is shorter since it coincides with the school break, but in the spring some of the circle sessions are replaced by interviews conducted in order to select students for special math classes at the eighth and ninth grade level. These interviews are not restricted to math circle participants: any interested student can come for an interview.

How do we run a circle session?

To begin with, all students get a handout with the main problems. Students read the problems and try to solve them. They can solve them in any order. An instructor might gently suggest that it could be a good idea to move down the list without spending too much time on any particular problem: if a problem takes too much time it might be better to leave it alone for a while and get back to it later.

A student who has a question or wants to discuss his or her solution simply raises a hand and an instructor comes by to talk. Students can explain their solutions orally, but it is useful to have at least some written notes with all necessary drawings and calculations. A conversation with an instructor is far from being a test. Sometimes a student has a solution but due to inexperience cannot articulate it, or he might totally misunderstand the problem, or he does not know some important facts, etc. The instructor's task is to help the student.

In the middle of the session there is a short break for anyone who needs it. Close to the break—either right before it or right after—solutions for the problems of the previous sessions are discussed with the whole class. The reason for doing it in the middle instead of the beginning of the session is to allow students to turn in those problems that they didn't finish in the previous meeting, but completed at home. While discussing old problems, some remarks concerning the current problem set might be appropriate; in some cases they might be offered even earlier if the topic seems to be hard for the students. It all depends on the head instructor, who might choose any moment to give a hint, or offer a funny problem to relax the class, or ask someone to come to the board and present her solution, or play a game.

Instructor explanations at the blackboard take only about 20 minutes out of a two-hour session. The rest of the time is devoted to students solving the problems and discussing them individually with instructors.

Students who are done with the main problems are provided with additional ones. Sometimes a student who, with plenty of time left, has solved all but one of the main problems and is stuck with the remaining one, might also be given additional problems, so as to be able to move on. At the end of the session the list of additional problems is given to anyone who asks.

There are always some students who solve very few or none of the problems, however easy the session might be. It is important that during any session an instructor comes to each student to give a hint or just to talk about the problems. A student who has not finished all the problems at the end of a session can solve them at home and discuss the solutions during the next session; however, there is no mandatory homework. Occasionally, hard additional problems are discussed several weeks after they were introduced, to give everyone a fair amount of time to work on them.

Each student's progress is registered in a special journal, where finished problems are marked with a plus sign. Children of this age love to compete: they are very happy to have many pluses, and they are disappointed to have just a few. We try not to focus children's attention on these signs because they are not the goal. It is the instructors who need these journals to measure the success of a session. It's better to register the results right during the class, since this gives a timely indication of which problems are harder, which require a hint, or pointing students into some particular direction.

Problem set structure: how problems are selected

The usual practice in regular schools is to create a lesson around a particular topic. For example, if today's topic is quadratic equations, then we explain some theory, give some examples, and solve some problems on quadratic equations. As a result, students quickly learn the new topic and can do standard exercises. This method definitely achieves results. However, I think that this method does not work well for math circles, especially for younger students.

One reason is that solving problems on the same topic for the whole session may be difficult and boring for a young student. Furthermore, if the instructor does not start a session with an explanation of how to solve problems on the new topic, a student who fails to figure out the key idea on his own can end up sitting through the whole session without solving a thing. On the other hand, if the instructor explains how to solve problems of a given type, it would leave less room for creativity, since it's enough just to remember what the teacher said and apply it to similar problems.

I choose a somewhat different approach. The progress is slower but, I believe, more reliable and interesting, and the method has some additional advantages. A typical handout looks like this:

The first problem is very easy. Sometimes it can be solved in different ways. One solution might not require any inventive thinking, just some straightforward work, while another might involve an interesting idea that would allow a student to solve the problem easily, quickly, and elegantly. For example, consider the following question: What is bigger, the sum of the first 50 odd natural numbers or the sum of the first 50 even natural numbers, and by how much? Some students will calculate both sums and compare them, but the answer can actually be made quite obvious without these cumbersome calculations. It is very useful to pose problems that have an intuitively obvious but wrong answer. For example: Is it possible for the product ab to be divisible by c^2 if neither a nor b is divisible by c? The "obvious" answer is, of course not. But in reality, ab can be divisible by c^2, or even by c^{100}. This problem makes an impression!

Actually, every problem set does have a particular topic, and approximately half of the problems are dedicated to it. Moreover, one medium-to-easy preliminary problem on this subject should be given in the previous handout, so that the students can think about this new type of problem beforehand. In the middle of the current session, when solutions of the previous handout are presented, this problem will be discussed. Thus, those students who have not yet figured out how to solve problems on the new topic will get a hint. The handout also contains some repetitive problems. I like asking the same questions dressed up differently. As an example, here are five questions that get at the same fundamental idea:

1. There are water taps in a school cafeteria. Each one can be open or closed. In how many ways can the water run in the cafeteria?

2. How many strings can one make of 0's and 1's, so that each string consists of 10 digits?

3. There are 10 apples growing on a tree. In how many ways can one pick some of them?

4. After school, 12 students decided to split into 2 groups, one to explore the city, and another to attend a class in computer science. In how many ways can they split?

5. The school cafeteria menu is always the same and consists of n different items. Peter wants to choose his breakfast differently every day; he can eat from 0 to n different items at a time. For how many days would he be able to do this?

Students often solve those problems as though they were completely new. But it makes me very happy if a student says "Wait! But we have already solved this problem!" This means that he has learned to see the essence of the problem and is not distracted by the appearance. Repeating problems that are essentially identical but presented in different formulations in a handout is also useful for those who missed some classes.

As already mentioned, each handout contains an easy, or medium-level problem, anticipating the topic of the next handout. One can give a problem on the same new topic in several subsequent sessions. Then the students themselves may gradually come up with a method to solve them. Later it will be easier and more natural for them to accept this method from the teacher when the topic is discussed in detail.

Sometimes students are given a whole sequence of problems, one per session, in which every problem helps to solve a later one. A good example is a sequence of problems about crossing the river or about catching a bus.

It is sometimes possible to give a hard, but captivating problem on a new topic to interest the students, but be sure to give them a lot of time for thinking. Then students will say at every new session, "Oh, let's discuss that problem — how can it be solved?" One can discuss the problem at the board together with the students, using brainstorming to obtain some intermediate result, and only discuss the complete solution several sessions later.

Additional problems based on either a new or an old topic are given to those who have mastered almost all of the main ones. They are designed for strong students and may be difficult or just interesting. They often require ingenuity and persistence to solve.

For younger students, it is good if many sessions have a mathematical game among the main problems. Kids can play it alone, with one another, or with an instructor. One can teach the children a lot of ideas with the help of the game problems. And what should one do if a student is persistently trying to submit a solution to a game problem based on a wrong strategy? Of course, one should play with him, but doing what? Beating him? That's not very good. One can give away the idea for the solution or disappoint the student. There is a wonderful technique that I learned from Nikolai Konstantinov. The instructor should adopt the strategy of the student and lose to him. If the student makes bad moves, the instructor may hint at the right ones and correct the wrong ones, but at the same time follow the strategy of the student at all times. As a result, the student wins, which is great, and at the same time is shown that his strategy is flawed.

Each session also has a geometric problem. They may be simple brain-teasers, cutting problems, or problems of classical geometry requiring only minimal knowledge.

Children will not be bored with these sessions. They can use their previously acquired knowledge to solve problems on the earlier topics, or they may try to do problems on the new topic, or may try to solve problems that just require ingenuity.

An important advantage of this approach is that students will learn how to choose the method according to the problem. As an example of the need for this, I had a case where a student was taking an exam on three topics:

induction, combinatorics, and integers. The student received a problem and worked on it for a while unsuccessfully and then said, "I can't solve this problem, because I can't figure out what topic it's about. Tell me what topic it's in so I can solve it." This was a bit depressing. I have to say that the student was in fact strong and is now successfully studying mathematics at a university.

It often does not occur to students that it is necessary to reason about the problem in order to solve it. In school, many students develop a habit of following a routine, often not understanding its meaning. One of my friends, university professor Gregory Rybnikov, told me that on a discrete mathematics exam he gave a reasonably good student the following nonstandard problem: "Prove that if $N > 1$, in any company of N people there will always be two who have the same number of friends in this company." For a long time, the student could not figure out how to approach this problem, and then the teacher gave him a hint. "Let's reason by contradiction: Let all the people in the company have a different number of friends. Note that each one in the company has no more than $N - 1$ friends." "I see," the student said, "the total number of people is N, and every one can have from 0 to $N - 1$ friends, so there is the same number N of possibilities. This means that someone has 0 friends, someone has one, someone has two, and so on until $N - 1$. Oh, this is an arithmetic progression! I can find its sum." The sum of an arithmetic progression has absolutely nothing to do with this problem, but the student, who had already almost solved the problem, lost it. He got used to following routines — if there is an arithmetic progression, one should probably find its sum.

At the circle children are surprised to find out that problems can be solved through reasoning. Sometimes someone is so surprised that a simple argument without formulas can solve the problem that he asks, "Can we argue this way? Is this correct? Is it rigorous?" This happens, for example, with Problem 2.1 on page 5, about figures made of squares.

The goal of the math circle is not to explore the problems of a particular type or master a lot of factual material, but to interest students in mathematics, to show that mathematics is a beautiful and interesting science, to teach them to reason, and to distinguish a solution from a nonsolution.

Not all students can attend the circle regularly, and skipping a few sessions won't impair their ability to work on a new handout. Although regular work is an important prerequisite in studying math, the most important one is probably interest. I know how to teach students who are interested — I can interest them even more — but there is almost nothing I can do with the indifferent ones. There are a lot of beautiful problems, and if the students can see that, they will succeed. To learn how to solve problems, one just needs to constantly work on them. Yuri Lysov, a student of Nikolai Konstantinov, believes that the meaning of a math circle is to show the students that they are able problem solvers.

Difficulties and Pitfalls

Various difficulties can arise. Inexperienced instructors sometimes cannot enforce discipline, and students talk too much and distract each other. Or a student may have been dragged into the circle by his parents, while he wants to run away and play soccer.

Sometimes the student-assistants cannot solve problems for sixth graders. It is very important that the problems be discussed with the instructors beforehand and that the instructors solve all the problems themselves, or at least learn the solutions for those they could not solve. The senior instructor should be able to solve all the problems and should have a great deal of knowledge and experience.

Students like asking questions, and this can lead to ones that are quite meaningful and difficult even if they are easy to formulate. The teacher also should not be afraid to say "I don't know" and find the answer by the next session. It can happen that the answer is too complex or even unknown. For example, it is unknown whether there is always a prime number in the interval from n^2 to $(n+1)^2$, and even the known fact that there is always a prime number in the interval from n to $2n$ is very difficult to prove.

Sometimes students visit the math circle after an Olympiad clamoring, "Tell us how to solve these problems." If that is the case, it is better to adjust the session's plan by including some of the problems from the Olympiad.

Students often infect each other with their enthusiasm, which is great, but sometimes there is not enough energy and the session is sluggish. It can happen that a problem set that appears accessible and interesting, at least for some students, is difficult or boring for others. When students find the problem set difficult, the head instructor must change the flow of the session and give out different problems. Another possibility is to start solving and discussing the problems together, dividing them into simpler parts, or giving hints. For example, if a game is played on a strip of size 1×100, the first piece of advice is to simplify the problem. What would be the answer if the strip has size 1×1, 1×2, or 1×3 instead? In finding answers to simpler problems, one can understand the general case. It is often very useful to look at special cases in order to try to guess the answer and then try to prove that the guess was correct. It is important that the students distinguish these intermediate guesses from a complete solution.

Sometimes the students, and even some teachers, believe that in each problem one must explain how they have found the solution. This is not always the case. For example, one might have a problem where one should come up with five numbers with a certain property. If the student just made a guess out of the blue but came up with five numbers with the desired property, this is wonderful: the ability to guess right is very useful in mathematics. Of course, to invent a general method for finding those numbers may also be important, and it is useful to talk about it, but that's no reason not to count this problem as solved. Sometimes the solution is just an inspired guess that comes from nowhere.

One should have a lot of patience in working with school children. A student might be initially inexperienced and understand very little, but it does not mean he is unable to learn mathematics. Once I had such a student, one who at first was only able to perform even the simplest tasks slowly and with great difficulty, but then decided to study seriously and worked all summer solving a whole collection of Olympiad problems. The student was admitted to a mathematics day school and became a successful mathematician.

One should not rush to cover a lot of topics and solve a lot of problems. If it is clear that the problems are too difficult, one must move more slowly, taking all the time that's needed.

Sometimes students who are not used to mathematics just do not understand the language we are using. Mathematics has a lot of conventions and tacit assumptions. Once I was trying to find of a problem in combinatorics for a specific student to solve, and came up with this: "How many ways are there to choose one person out of ten people?" The student's answer, 9, disconcerted me. "Why nine?" I asked. "I would not choose myself," replied the student. We were talking in different languages; he understood the simplest formulation in his own completely different way.

What a joy it is when students discover an insight for themselves. For example, one student tackled this problem: how many pairs can be made of two digits if the first digit in a pair can be any digit from 0 to 9, and the second digit in a pair can be any digit from 0 to 9? He started to write down the pairs. First he wrote down all pairs where both digits are the same, and so gradually got the answer 99 (he forgot the pair 00). That answer simply shocked him. For a while he was looking frantically at his notes. Then he exclaimed, "It's just all numbers from 1 to 99!" This student made a small mathematical discovery, and this is important, even if it took him a long time to come to it, and despite making a small mistake in the process.

Not everyone will become interested in mathematics, but if a student has the desire and perseverance, something good is bound to happen. It is also worth mentioning that this book contains problem sets of one particular math circle for one particular year. I changed almost nothing, although some sets turned out to be more complicated and some easier than planned. Lessons vary from year to year and from class to class. This compilation may be of great help to the instructor, but, of course, one should not just copy these lessons. Depending on the students, it may be necessary to change the topics, difficulty level, number of problems, etc. It is all in the hands of the teacher!

What's next?

At the end of the year, or at the end of a two- or three-year long course, students sometimes receive humorous "little math scholar" certificates. Attending the circle does not bring any formal benefits, but the acquired knowledge and skills help them a lot more than any certificates. There are some cases, however, when a document can come in handy. Once I was approached by a sixth grader who complained that he could not enroll in a particular very good library, because, they said, he was too young. The boy asked me for help. I wrote a letter, stating that the boy was a good student at "The Little MechMat" and asked for the student to be allowed to enroll in the library. An official paper with a Moscow State University seal did the job, and the boy was enrolled.

School children who finish one year of the circle may transfer to its continuation the next year. The instructors of The Little MechMat usually also transfer to the next grade and some of them organize groups for the continuing students in parallel with groups for newcomers and for those who did not have too much progress over the past year.

Many students get into mathematics day schools and start studying the subject more intensively. Of the students in these schools, sometimes half are former participants in math circles. The main education in math schools takes place at the formal lessons where attendance is mandatory. Yet it is important even in a mathematics magnet school to have lessons that students can attend at will. For all students of my math class who are interested, I lead a separate math circle, and a seminar based on the journal *Kvant*.

A mathematical circle in a mathematical school is usually attended by a few strong students since the regular school load is already difficult enough. The sessions are very informal. Often I just bring a collection of Olympiad problems and, right in the class, choose the problems that I like. The students can participate in the selection. The teacher can make a session on some topic, or introduce some interesting idea to the students. Sometimes, students themselves bring problems, and we solve them. Fans of geometry shout "Give us an interesting problem in geometry." Solutions are discussed with students individually at first, and then each one can present his solution at the board. Sometimes, if the problem is difficult, we solve it together. This is an exciting time for the children. After the circle we have tea with cookies.

Since I work in *Kvant*, I lead a seminar affiliated with this journal. Articles sent to the journal are subject to refereeing, but the referees are not only math experts—and one of the referees is a student. The student must understand the article and give a talk. At the time of the talk, which is attended by teachers and students, the speaker will answer questions from other students. We often think together on a difficult point. The seminar resembles a research seminar in that it is sometimes difficult to understand what the author had in mind. We sometimes find typos, errors, or obscure passages. We write a review for the author with recommendations on how to improve the paper. This is very important information, coming from prospective readers. The children see that they are doing important work; they determine whether an article will be published. Later, when they see the article in the journal, they feel quite encouraged. Sometimes there is even a dispute as to who will give a talk in the next seminar. Often there are too many students willing to give one.

Besides the obvious benefits for the journal, there is a tremendous advantage for the students as they learn how to read mathematical papers and understand them on their own. This is something that is lacking in a regular classroom. Students learn to speak and learn to listen. The seminar ends with the tea and further informal discussions. Not all students can give a good talk, but all try and gradually learn. In fact, the talks of some students are better than many of the lectures by university professors. A teacher may also base a seminar on old papers, which I often do. From 1990 through 2001, the journal *Quantum*, the American brother of *Kvant*, was published by the National Science Teachers Association and Springer-Verlag, and it can serve as a very good source for such work.

Where do the problems in this book come from?

Almost none of the problems in this book were invented by me; they were selected from a variety of sources, including various Olympiad and math circle collections, books, journals, and sometimes the oral folklore. Many problems entered the folklore so long ago that it does not seem possible to name the authors. My contribution is that I selected them, compiled them into the circle's sessions, and wrote solutions.

I hope that this book will be interesting to students who love math and are not afraid of difficulties, and maybe even help someone to fall in love with mathematics.

Acknowledgments

I would like to express my gratitude to all who, in one way or other, contributed to the preparation of this book, and especially my student Andrey Ionov, who read the original manuscript and made many comments; the brilliant artist Eduard Nazarov, who allowed the use throughout the book of his lovely cartoons, formerly published in *Kvant*; Alexey Voropaev, who suggested an English translation for the YumYum poem of Problem 6.5; Eugene Epifanov, who made some of the geometric drawings using Metapost; and Jonathan Borofsky, for the use of his art in the Mathematical Maze.

Without the dedication of Tatiana Shubin, who translated the manuscript, of David Scott, who edited the translation, and of Silvio Levy, who handled typesetting, layout, and final editing, this book would never have seen the light of day. All three made valuable suggestions, and their contribution far exceeds what is expected or translators and editors. A huge thank you to them.

Problem Set 0

Problem 0.1. Grandma takes five minutes to climb from the first floor of a building to the fifth. If she climbs at the same speed, how long will it take her to get to the ninth floor from the first?

Problem 0.2. Juan and Candice are using a spring scale to weigh their book bags. When they are weighed separately, the scale shows 3 lbs and 2 lbs. When they weigh them together, the scale shows 6 lbs.

"That can't be right," said Candice. "Two plus three doesn't equal six!"

"Don't you see?" answered Juan. "The pointer on the scale is off."

How much do the book bags actually weigh?

Problem 0.3. Use the fingers of one hand to count as follows: the thumb is first, the index finger is second, and so on to the pinkie which is fifth. Then reverse direction so that the ring finger is sixth, the middle finger is seventh, the index finger is eighth, and the thumb is ninth. Reverse direction again back toward the pinkie with the index finger tenth, and so on. If you continue to count back and forth along the fingers of one hand, which finger will be the 1000th one?

Problem 0.4. A dot is marked in a circle. (a) Cut the circle into at most three parts so that, by rearranging the parts, you get a circle with its center at the dot. (b) Is it possible to do so by cutting the circle into at most two parts?

Problem 0.5. A brother leaves his house 5 minutes after his sister. If he walks at 1.5 times her speed, how long will it take him to catch up?

Problem 0.6. The diagram shows the rolling track of a bulldozer, seen from the side. The bottom is in contact with the

ground. If the bulldozer moves forward 10 cm, how many centimeters does the point marked A move?

Problem 0.7. Together Winnie the Pooh, Owl, Rabbit, and Piglet ate 70 bananas. Each ate a whole number of bananas, and each ate at least one. Pooh ate more than each of the others; Owl and Rabbit together ate 45 bananas. How many bananas did Piglet eat?

Additional Problems

Problem 0.8. While walking in the park, Nicole and Valerie came to a large round clearing surrounded by a ring of cottonwood trees and decided to count the trees. Nicole walked around the clearing and counted all the trees. Valerie did the same, but started from a different tree. Nicole's 20th tree was Valerie's 7th, while Nicole's 7th tree was Valerie's 94th. How many trees were growing around the clearing?

Problem 0.9. A group from a summer camp leaves a forest where they have been gathering flowers. They walk in boy-girl pairs, and in each pair the boy has either three times as many or one third as many flowers as the girl has. Is it possible for the whole group to have 2006 flowers?

Problem Set 1

Problem 1.1. A beaker filled to the brim with water weighs 5 pounds, while the same beaker filled halfway weighs 3.25 pounds. How many pounds of water can the beaker hold?

Problem 1.2. Which is greater, 333333×444444 or 222222×666667? By how much?

Problem 1.3. Given a triangle ABC with angle $B = 90°$ and $AB = BC = 1$, and a point M chosen at random on AC, is it possible to tell what the sum of the distances from M to AB and from M to BC will be?

Problem 1.4. When Peter broke his piggy bank, it contained no more than 100 coins. He divided the coins into piles of 2 coins each, but was left with one extra coin. The same happened when Peter divided the coins into piles of 3 coins, piles of 4 coins, and piles of  5 coins. Each time he was left with one extra coin. How many coins were in the piggy bank?

Problem 1.5. A rectangular steel plate measuring 17×10 inches has been traced on a piece of paper. Using only the plate, the paper, and a pencil, find the center of the paper rectangle.

Problem 1.6. Angela has 7 potatoes, Minh has 5, and Greg has none. They combine their potatoes to make a bowl of mashed potatoes and share the bowl of potatoes equally among the three of them. In exchange for his share, Greg gives Minh and Angela 12 pieces of chocolate. How should they divide the chocolate between them, if they are to be fair?

Problem 1.7. Is it possible to cut several circles out of a square of side 10 cm, so that the sum of the diameters of the circles would be 5 meters or more?

Additional Problems

Problem 1.8. For an experiment a researcher puts a dot of invisible ink on a piece of paper and also draws a square with regular ink on the paper. In the experiment, a subject will draw a visible straight line on the page and the researcher, who has on special eyeglasses for spotting the dot, will tell the subject which side of the line the dot of invisible ink is on. If the dot is on the line, the researcher will tell the subject it is on the line. What is the smallest number of straight lines the subject needs to draw to figure out for sure whether the invisible dot lies in the square?

Problem 1.9. A train moves in one direction for 5.5 hours. If the train covers any 100-mile segment of the journey in 1 hour: (a) Is the train necessarily moving at a constant rate? (b) Is the train's average speed necessarily 100 mph?

Problem Set 2

Problem 2.1. Here is a series of figures:

The first consists of one square. How many squares are in the 100th figure? How many squares are in the first 100 figures altogether?

Problem 2.2. A grasshopper jumping along a straight line can jump 6 or 8 inches in either direction. Can it reach a point that is (a) 1.5 inches away from its original position; (b) 7 inches away; (c) 4 inches away?

Problem 2.3. A cardboard rectangle with area 1 is cut into two pieces along a line segment that connects the midpoints of two adjacent sides. Find the areas of the two pieces.

Problem 2.4. Ten gnomes are playing checkers. Each plays one game with each of the others. (a) How many games did each gnome play? (b) How many games were played in total?

Problem 2.5. The windows on a subway train are as shown in the drawing. The curves forming the rounded corners are all arcs of a circle. A portion of the window is opened 10 inches. The height of the open section is 25 inches. Find the area of the opening.

Problem 2.6. Suppose in a given collection of 2002 integers, the sum of any 100 of them is positive. Prove that the sum of all 2002 of the integers is positive.

Problem 2.7. An ant is sitting in a corner of the floor of a cubical room. It wants to move to the opposite corner using the shortest route. It can only move along the walls, floor, and ceiling of the room. What path should it take?

Additional Problems

Problem 2.8. A bus, a truck, and a motorcycle pass a stationary observer at equal time intervals. They pass another observer farther down the road at the same equal time intervals, but in a different order. This time the order is bus, motorcycle, truck. Find the speed of the bus, if the speed of the truck is 30 mph, and the speed of the motorcycle is 60 mph.

Problem 2.9. The numbers 2^{2002} and 5^{2002} are expanded and their digits are written out consecutively on one page. How many total digits are on the page?

Problem Set 3

Problem 3.1. (a) There are 10 baskets arranged in a circle. Is it possible to arrange several oranges in the baskets so that the difference in the number of oranges in any two adjacent baskets will be 1? (b) What if there are three baskets? (c) What if there are nine baskets?

Problem 3.2. Which is greater, the sum of all even numbers from 0 to 100 or the sum of all the odd numbers from 1 to 99? By how much?

Problem 3.3. (a) A game board is made up of 25 squares arranged in a 1×25 rectangle. In playing the game, two players, Alice and Bob, alternate moving a marker either 1 or 2 spaces forward, with Alice moving first. The marker is initially in a square at one end of the board and always moves toward the other end. The player who cannot make a move that stays on the board loses. Which of the two players can ensure victory? (b) What if the rules are changed so that the marker can be moved 1, 2, or 3 spaces forward?

Problem 3.4. "Andy has more than 1000 books." "No, he has less than 1000 books." "Well, he has to have at least one book." If it is known that only one of these three statements is true, how many books can Andy have?

Problem 3.5. (a) Out of nine identical looking coins, one is counterfeit, and is lighter than the others. How can you discover the fake with only two weighings with a two-pan balance? (b) Find the least number of weighings on the balance necessary to discover one counterfeit among 27 coins.

Problem 3.6. (a) The director of a secret service has put together a list of mutual surveillance assignments for his seven agents, codenamed 001 through 007. Agent 001 will watch the agent who is watching 002; agent 002 will watch the agent who watches 003; and so on. Agent 007 will watch the agent watching 001. Can you deduce who watches whom? (b) Would it be possible to carry out similar instructions with 8 agents?

Problem 3.7. (a) Draw a closed 6-segment broken line so that each segment intersects only one of the remaining segments in a point other than an endpoint of a segment. (b) Is a similar figure with 7 segments possible?

Additional Problems

Problem 3.8. Jack and Jill are playing a game. Out of a common pile of 777 matches, each can remove either 7 or 77 matches each time it is their turn. They alternate turns removing matches with Jack going first. The winner is the player who causes his opponent to be unable to make further moves. Who wins?

Problem 3.9. A hallway is completely covered by several rectangular rugs, each the width of the hallway. Some rugs may overlap. (a) Prove that it is possible to remove several rugs without changing the position of the rugs not removed so that every portion of the hall is still covered, but no more than two rugs overlap at any point. (b) Prove that it is possible to remove still more rugs so that the remaining rugs will not overlap at all and will cover at least half of the hallway.

Problem 3.10. In a herd of 101 cows, each weighs a whole number of pounds. If any one cow is removed from the herd, the remaining cows can be divided into two groups of 50 cows each with the total weight of all cows in the first group equal to the total weight of all the cows in the second group. Prove that all of the cows weigh the same.

Problem 3.11. Each of the letters F, I, V, E in this multiplication stands for a different digit:

$$\begin{array}{r} \text{FIVE} \\ \times\ \text{FIVE} \\ \hline \ast\ast\ast\ast\text{F} \\ \ast\ast\ast\ast\text{I} \\ \ast\ast\ast\ast\text{V} \\ \ast\ast\ast\ast\text{E} \\ \hline \ast\ast\ast\ast\ast\ast\ast\ast \end{array}$$

What are the values of the letters?

Problem Set 4

Problem 4.1. In a certain small country there are 4 cities: A, B, C, and D. There are 5 roads between A and B, 4 roads between B and C, 2 roads between A and D, and 3 roads between D and C. In how many 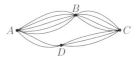 different ways is it possible to travel (a) from A to C via B; (b) from A to C via B or D?

Problem 4.2. (a) If you have a 7-minute hourglass and an 11-minute hourglass, can you use them to boil and egg for 15 minutes? (b) Some have solved (a) by letting an hourglass run before putting the egg in the boiling water. Is it possible to start the egg boiling at the same time you start an hourglass and still boil the egg for 15 minutes?

Problem 4.3. In a given nation, every 20th mathematician is also a musician, while every 30th musician is also a mathematician. Are there more mathematicians or musicians in the nation? How many times more?

Problem 4.4. (a) There are three bulbs in a very short string of Christmas lights; one is red, another blue, and the third is green. Each bulb can be either on or off. In how many different ways can the string light up? (b) What if the string is made up of five different bulbs of five different colors?

Problem 4.5. Check the validity of the equations in the picture and try to prove the following statement: For any positive integer n, we have $1 + 2 + \cdots + (n-1) + n + (n-1) + \cdots + 2 + 1 = n^2$.
(The balloon next to the pyramid contains a hint.)

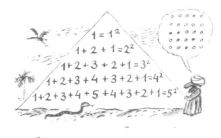

Problem 4.6. Strictly speaking, an anagram of a word or phrase is a rearrangement of the letters to form a different word or phrase. In anagrams, spaces are usually ignored. For example one anagram of "anagram" is "nag a ram". In mathematics, and for this problem, we often use "anagram" to mean any permutation of letters in a word, and therefore would consider "aaarngm" an anagram of "anagram". (a) How many anagrams does the "word" REALSPY have? (b) Can you find one that means an herb? One that describes the members of a team? Can you find any others that are

English words? (c) Decipher the following sentence where the correct words are replaced by their anagrams: VOLES ATTACHELAMMI BERMSLOP AYLID. (d) How many anagrams does the word APPLE have? (e) How many does BAOBAB have?

Problem 4.7. (a) Into how many parts can two distinct lines divide a plane? Draw an example of every possible case. (b) Into how many parts can three distinct lines divide the plane? Draw an example of every possible case. (c) What if there are four lines?

Additional Problems

Problem 4.8. A page in a calendar is partially covered by the preceding page as in the picture. Which has greater area, the covered part or the uncovered part?

Problem 4.9. Is it possible to put beans on the squares of an 8×8 grid so there are the same number of beans in any two columns and a different number of beans in any two rows?

Problem 4.10. (a) A toy factory produces multicolored triangular pyramids. Each pyramid has four equilateral triangular faces with one yellow, one red, one blue, and one green. How many different coloring patterns can the factory produce on these triangular pyramids? (b) Consider the same problem if the factory produces cubes each having square faces with one yellow, one red, one blue, one green, one white, and one black.

Problem Set 5

Problem 5.1. All four sides of a quadrilateral are congruent. Is it necessarily a square?

Problem 5.2. There are 25 students in a class. In how many different ways is it possible to choose the following groups from the class? (a) A hall monitor and the president. (b) Two hall monitors. (c) Three hall monitors.

Problem 5.3. How can you cut 27 inches from a 144-inch ribbon without a ruler?

Problem 5.4. A group of 15 children have gathered 100 nuts. Prove that 2 of them have gathered an equal number of nuts.

Problem 5.5. (a) A company has 67 employees. Of them, 47 speak Spanish, 35 speak German, and 23 speak both. How many employees don't speak either language? (b) Suppose that, in addition, 20 employees at the same company speak French, 12 employees speak French and Spanish, 11 speak French and German, and 5 speak all three. How many employees don't speak any of the three languages?

Problem 5.6. (a) Is it possible to connect 7 light bulbs to a power source using only 3 switches so that it would be possible to light up any number of the bulbs, from 0 to 7? (b) What if there are 8 bulbs and 3 switches?

Problem 5.7. Two concrete staircases, both one meter high and two meters long, are covered with runners (long strips of cloth 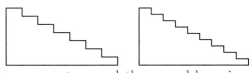 or carpeting). The first staircase has seven steps and the second has nine, as shown in the diagram. Will a runner that completely covers the first staircase also completely cover the second?

Additional Problems

Problem 5.8. Two mercury thermometers are hung next to each other as shown on the right. At what temperature will the mercury in both thermometers be at the same height?

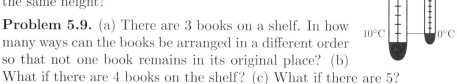

Problem 5.9. (a) There are 3 books on a shelf. In how many ways can the books be arranged in a different order so that not one book remains in its original place? (b) What if there are 4 books on the shelf? (c) What if there are 5?

Problem 5.10. From the set of numbers 1, 2,..., 50, twenty-six numbers are chosen. Prove that one of the numbers is divisible by another.

Problem Set 6

Problem 6.1. One of the diagonals of a rhombus is equal to a side. What are the measures of the angles of the rhombus?

Problem 6.2. Which is greater, $\frac{2005}{2006}$ or $\frac{2006}{2007}$?

Problem 6.3. Can the square of an integer end with 2?

Problem 6.4. "We have 25 students in our class, and each is friends with exactly seven classmates." Why can't that statement be true?

Problem 6.5. The YumYum language is written in a mixture of Russian and English characters. Here's a little poem in YumYum with the translation of each verse:

Ам ту му ям.	=	Giant cat was scary.
Ту ля бу ам.	=	Mouse saw giant cat.
Гу ля ту ям.	=	Scary cat ate mouse.

Based on this, construct a fragment of a YumYum-English dictionary.

Problem 6.6. Prove that if you have a collection of 11 natural numbers, there must be two of those numbers whose difference is divisible by 10.

Problem 6.7. Carol is on an airplane flight. First she read a book; then she slept; then looked out of the window, and then drank orange juice. Each of these activities except the first took exactly half as long as the previous one. She started reading her book at noon and finished her orange juice at 1 p.m. When did Carol start looking out of the window?

Problem 6.8. More than half of the surface of the spherical planet Urth is land. Assuming Urth is solid beneath its surface, prove it is possible to dig a straight tunnel through the center of Urth and connect land with land.

Additional Problems

Problem 6.9. In a set of many 2×1 rectangles of the same size, some are blank and some have a diagonal, as in the diagram. Eighteen rectangles will be chosen and assembled into a 6×6 square in which no diagonals on the rectangles meet. What is the least number of blank rectangles needed to do this?

Problem 6.10. How many 15-digit strings of 0's and 1's are there that do not have two 0's next to each other?

Problem 6.11. Can an equilateral triangle be covered by two smaller equilateral triangles?

Problem Set 7

Problem 7.1. A train that is 180 meters long passes a signal in 90 seconds. How long will it take it to cross a bridge that is 360 meters long?

Problem 7.2. Four posts have been placed at the corners of a square pond. How can the pond be expanded without removing the posts so the area doubles, the shape remains square, and the posts don't end up in the water?

Problem 7.3. There are 25 students in a class. (a) Prove that two students have birthdays in the same month. (b) Are there necessarily three such students?

Problem 7.4. Find the remainder when: (a) 3^{100} is divided by 5; (b) 5^{100} is divided by 3.

Problem 7.5. Suppose that 26 numbers are chosen from the set $1, 2, \ldots, 50$. Must two of the numbers differ by 1?

Problem 7.6. Find the areas of the figures in the following diagrams, if the area of each square is 1.

Problem 7.7. (a) How many different 10-digit numbers can be written using only the digits 1 and 2? (b) There are ten mangoes in a tree. How many ways are there to pick several of them?

Problem 7.8. A 200×3 rectangle is drawn on graph paper along the grid lines. How many grid squares will a diagonal of the rectangle cross?

Problem 7.9. Prove that in a class with at least two students there are at least two students with an equal number of friends in the class.

Additional Problems

Problem 7.10. A white rook and a black bishop are on a nonstandard chess board. They take turns moving according to standard chess rules: the bishop moves diagonally any number of squares, and the rook moves up, down or across any number of squares. The rook moves first. How should the rook move to capture the bishop if the size of the board is: (a) 3×10; (b) 3×1000?

Problem 7.11. A grasshopper is sitting at the edge of a circular patio of diameter 3 meters. Every hop the grasshopper takes is exactly 2 meters long. If the grasshopper never leaves the patio, what points of the patio can it reach?

Problem 7.12. There are 2003 representatives of four Middle-Earth races sitting around a round table: humans, dwarves, goblins, and elves. Humans never sit next to goblins, and elves never sit next to dwarves. Prove that at least one pair of representatives sitting next to each other are from the same race.

Problem Set 8

Problem 8.1. Cut the first triangle in the diagram into parts that can be reassembled to form the second triangle.

Problem 8.2. A certain type of bacteria multiplies so fast that the area they cover in their Petri dish doubles every ten minutes. One bacterium was placed in a Petri dish, and the dish became completely covered with bacteria in exactly 5 hours. How long will it take for the dish to get completely covered if we start with two bacteria?

Problem 8.3. Fifteen black dots have been marked on a square of white paper of side 4 cm. Prove that it is possible to cut out a square of side 1 cm from the larger paper so that the smaller square will not contain any dots inside.

Problem 8.4. A flock of white geese flew over a string of lakes. As they came to each lake, half of the remaining geese plus half a goose landed on that lake, while the others flew on. By the seventh lake, all the geese had landed. How many geese were in the flock?

Problem 8.5. On a sheet of paper draw (a) 4 dots, (b) 5 dots, (c) 6 dots, so that any 3 dots will be vertices of an isosceles triangle.

Problem 8.6. On a stormy night, ten guests came to a dinner party and left their shoes in the foyer to keep the carpet clean. After dinner there was a blackout, and the guests, leaving one by one, put on, at random, any pair of shoes big enough for their feet. (Each pair of shoes stays together.) Any guest who couldn't find a pair of shoes big enough spent the night. What is the largest number of guests who might have had to spend the night?

Problem 8.7. Two mirrors are placed parallel to each other as in the diagram on the right. A ray of light from point A hits the first mirror at point B and eventually reaches point C. Reflections from each mirror follow the usual law: The angle of incidence is equal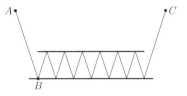
to the angle of reflection. Would the ray reach point C if the distance between the mirrors is doubled by raising the top mirror? If so, how will the length of the light's path from A to C change?

Problem 8.8. A $3'' \times 3'' \times 3''$ cube is to be cut into twenty-seven $1'' \times 1'' \times 1''$ cubes. What is the least number of cuts needed to do this? You are allowed to rearrange the pieces after each cut. Assume each cut is parallel to a face of the cube.

Additional Problems

Problem 8.9. Are there triangles that can be cut into: (a) three congruent triangular pieces; (b) four congruent triangular pieces; (c) five congruent triangular pieces.

Problem 8.10. A Xork lands on a planet that contains 100 Yorks. Every day after that, a battle takes place where each Xork destroys one York; right after the battle, every Xork and every York split into two. Prove that, sooner or later, all Yorks will be destroyed. How long will it take for that to happen?

Problem 8.11. (a) Place seven stars in a 4×4 grid so that, no matter which two rows and which two columns are erased, at least one star will remain. (b) Prove that if six stars are placed in a 4×4 grid, one can always erase all of them by erasing two rows and two columns.

Problem Set 9

Problem 9.1. Which is larger, $1 + 2 + 4 + 8 + 16 + 32 + 64 + 128 + 256 + 512 + 1024$ or 2048? By how much?

Problem 9.2. Each day at noon, a steamship leaves Savannah for Belfast while a steamship of the same line leaves Belfast for Savannah. Each ship spends exactly seven 24-hour days at sea, and all travel along the same route. How many Savannah to Belfast ships will a Belfast to Savannah ship meet while underway?

Problem 9.3. The length of one side of a triangle is 3.8 inches, and the length of another side is 0.6 inches. If the third side is a whole number of inches, find its length.

Problem 9.4. Simplify the fraction
$$\frac{1 \cdot 2 \cdot 3 + 2 \cdot 4 \cdot 6 + 4 \cdot 8 \cdot 12 + 7 \cdot 14 \cdot 21}{1 \cdot 3 \cdot 5 + 2 \cdot 6 \cdot 10 + 4 \cdot 12 \cdot 20 + 7 \cdot 21 \cdot 35}.$$

Problem 9.5. Using a pencil, an unmarked ruler, and a sheet of graph paper, how can you draw a square with area (a) double; (b) 5 times larger than the area of one square of the grid?

Problem 9.6. Which is greater, the sum of the lengths of the sides of a quadrilateral, or the sum of the lengths of its diagonals?

Problem 9.7. The menu in a school cafeteria always has the same 10 different items. To vary his meals, George decides to buy a different selection for every lunch. He can eat anywhere from 0 to 10 different items for lunch. (a) For how many days can he eat without repeating a selection? (b) What is the total number of items he will eat in that time?

Problem 9.8. Is it possible to write more than 50 different two-digit numbers on a blackboard without having two numbers on the board whose sum is 100?

Additional Problems

Problem 9.9. Show that the area of the green region of the regular pentagonal star in the picture is exactly half of the total area.

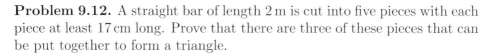

Problem 9.10. Find the sum

$$6 + 66 + 666 + 6666 + 66666 + \cdots + 66\ldots 66$$

if the last string of 6's has 100 digits.

Problem 9.11. Is it possible to find a number of the form $11\ldots1100\ldots00$ that is divisible by 2003?

Problem 9.12. A straight bar of length 2 m is cut into five pieces with each piece at least 17 cm long. Prove that there are three of these pieces that can be put together to form a triangle.

Problem Set 10

Problem 10.1. Piglet, Eeyore, and Winnie the Pooh live in houses connected with straight paths that form a triangle. In doing his exercises, Piglet ran from his house to Eeyore's, then to Pooh's, then home. At the same time, the thoughtful Pooh walked to Eeyore's house and back. Whose path was longer?

Problem 10.2. Is it possible to measure exactly 4 liters of water using a faucet, one 3-liter beaker and one 5-liter beaker?

Problem 10.3. A merchant brought a bag of nuts to sell at the market. The first customer bought 1 nut; the second 2 nuts; the third 4 nuts, and so on with each customer buying twice as many nuts as the previous one. The nuts the last customer bought weighed 50 lbs. The merchant had one nut left over. If all the nuts weighed the same, how much did the merchant's bag weigh at the beginning of the day?

Problem 10.4. True or false?
(a) Any dollar amount, beginning with $4, can be paid with $2 and $5 bills.
(b) Any tenge amount, beginning with 8, can be paid with 3-tenge and 5-tenge coins.

Problem 10.5. Two towns are near a straight highway. Where on the highway should a joint bus stop be located so that the sum of the distances from it to the towns will be the least if the towns are: (a) on opposite sides of the highway; (b) on the same side of the highway?

Problem 10.6. Winnie the Pooh and Piglet go to visit each other. They leave at the same time and each walks at a steady pace, but, because they are preoccupied with counting blackbirds that are flying by, they don't notice each other when they meet, and they end up at each other's houses. Piglet gets to Pooh's house 4 minutes after they passed each other, and Pooh gets to Piglet's house 1 minute after they passed each other. How long did each take to get to the other's house?

Problem 10.7. An 8×8 grid is covered by 1×2 dominoes. Prove that two of the dominoes form a 2×2 square.

Problem 10.8. Given a convex quadrilateral, find a point inside it for which the sum of the distances to each vertex is the least.

Problem 10.9. Suppose there is a group of pirates who want to divide their treasure. Each one of them is sure that he would divide the treasure equally, but none of the others trust him. What could the pirates do to divide their treasure in a way that, after the division, each would be sure that he received at least an equal share of the treasure if: (a) there are two pirates in the group; (b) there are three pirates in the group; (c) there are four pirates in the group; (d) there are k pirates in the group.

Additional Problems

Problem 10.10. A grasshopper is sitting in the corner of a square patio that is 2 meters on a side. Each jump the grasshopper makes covers exactly 2 meters. What points can it reach if it never leaves the patio?

Problem 10.11. Fifteen 2×2 squares have been cut from an 11×11 sheet of graph paper. All the cuts were made along grid lines. Prove that one more such square can be cut from the remaining paper.

Problem Set 11

Problem 11.1. To check if a piece of cloth is square, a tailor folds it across each of its diagonals and checks to see if the edges meet. Is this enough?

Problem 11.2. Can the sum of four consecutive natural numbers ever be divisible by 4?

Problem 11.3. (a) Two people take turns picking stones from two piles each with 9 stones. Each player can take any number of stones from one pile per turn. The winner is the last one to take a stone. Which of the players, the one who goes first or second, can always win, and how should he play to do so? (b) What if there are 3 piles of 9 stones? (c) What if there are 4 piles?

Problem 11.4. A piece of paper can be ripped into either 4 or 6 pieces. Prove that, with these two rules, it is possible to rip the paper into 9 or any larger number of pieces.

Problem 11.5. A logging company wants to chop down a forest that is 99% pine trees, but the Forest Service has objections. The logging company then suggests it will only cut pines, and, after they cut, the forest will be 98% pines. What part of the forest will be chopped down?

Problem 11.6. Is it true that if $n \geq 6$, a square can always be cut into n squares? We are not assuming that the smaller squares are all the same size.

Problem 11.7. Two towns, A and B, are on opposite sides of a canal with parallel straight sides. A road with a bridge that crosses the canal perpendicularly is to be built. Where should the bridge be located so that the length of the road from A to B is minimized?

Problem 11.8. A cue ball is shot out of the corner of a pool table at a 45 degree angle. Which corner of the pool table will the ball reach first, and how many times will it hit the sides until it does so if: (a) The pool table is 3×5; (b) The table is 3×1001?

Treat the cue ball as a point and assume it rebounds from the walls according to the familiar rule that the angle of incidence is equal to the angle of reflection.

Additional Problems

Problem 11.9. Several straight lines drawn on a sheet of paper divide it into polygons. Is it always possible to color each polygon with one of two colors so that any two polygons that share an edge are different colors?

Problem 11.10. (a) Is it possible to construct a decagon from any 10 straight rods? (b) If three quadrilaterals have been constructed from 12 straight rods, is it possible to construct four triangles from these 12 straight rods? (c) If four triangles have been constructed from 12 straight rods, is it possible to construct three quadrilaterals from these 12 straight rods?

Problem 11.11. Prove that the numbers $16, 1156, 111556, 11115556, \ldots$ are all perfect squares.

Problem Set 12

Problem 12.1. Twenty points are placed on a sheet of paper, and a line is drawn through every two points. What is the least and what is the greatest number of different lines that might be formed?

Problem 12.2. A chunk of consecutively numbered pages has fallen out of a folder. The first page of the chunk has number 463, and the last has the same digits but in a different order. How many sheets of paper were dropped? (Each sheet is two pages with consecutive numbers.)

Problem 12.3. Mike rips a sheet of newspaper into 8 pieces. He rips one of the resulting pieces into 8, and so on. Can he rip the paper into 2002 pieces?

Problem 12.4. A marker has been placed in the lower left corner of a 9×9 board. Two players take turns choosing a direction, right or up, and the number of spaces to move the marker in that direction. The winner is the last to make a move. Which of the players, first or second to move, can always win and what is the strategy to do so?

Problem 12.5. Prove that an equilateral triangle can be cut into six or any greater number of equilateral triangles. The triangles need not be congruent.

Problem 12.6. Sam and Alex live in the same apartment building and leave for school at the same time. Each of Sam's steps is 10% longer than Alex's, but Sam takes 10% fewer steps per minute than Alex. Who will get to school first?

Problem 12.7. (a) Two mirrors form a 30-degree angle. A light beam enters this angle parallel to one of its sides and is reflected from the sides according to the usual law that the angle of incidence is equal to the angle of reflection. Prove that the beam will eventually leave the angle. How many times will it reflect off of the mirrors before leaving? (b) What if the angle between the mirrors is 20 degrees? (c) What if it is 50 degrees?

Problem 12.8. Prove that each natural number can be written as the sum of several different powers of two. That is, any natural number can be written as a sum of numbers from the set $\{1, 2, 4, 8, 16, \dots\}$ with each power of 2 used at most once. For example: $100 = 64 + 32 + 4$.

Additional Problems

Problem 12.9. The numbers 1 trough 16 are written in a table as in the diagram. A plus sign or a minus sign is written in front of each number so that there are two pluses and two minuses in each row and column. Prove that the sum of the resulting numbers is 0.

1	2	3	4
5	6	7	8
9	10	11	12
13	14	15	16

Problem 12.10. Fifty points have been marked on a sheet of paper. Is it always possible to draw a line that separates the points into two groups of 25 points each?

Problem 12.11. Is it possible to find 10 different numbers such that the product of any two of them is divisible by the sum of all 10 numbers?

Problem Set 13

Today's problem set deals with goats. Goats are ravenous and consume everything they can reach. Because of this, they are usually kept on a rope.[1]

Problem 13.1. Draw the section of a pasture consumed by a goat if the goat is tied to a single stake planted in the pasture.

Problem 13.2. A mathematician took a walk on a field holding a goat on a 1-meter-long rope. The mathematician's path was rectangular with dimensions 3 meters by 5 meters. Draw the section of the field the goat will have consumed by the end of the mathematician's walk.

Problem 13.3. How can a goat be constrained to an eye-shaped field: In other words, how can the goat be tied using ropes and stakes, so that it can eat only the grass in the field?

Problem 13.4. A rope has been stretched between two stakes in a field. A goat is tied to this rope with another rope that is free to slide along the first rope. What is the shape of the portion of the field the goat can eat?

Problem 13.5. How can a goat be constrained to a field in the shape of a (a) semicircle; (b) square? (c) rectangle?

Problem 13.6. How can a goat be constrained to a field in the shape of a (a) triangle; (b) regular hexagon?

Problem 13.7. Dogs can be used to herd goats because a goat will not occupy a space that a dog can reach. However don't let a dog run free since it will chase the goat constantly, and never let the goat rest or eat. (a) How can one dog hold a goat in a ring? (b) What about in a semicircle? (c) Using dogs, contain an untied goat in a triangle.

Problem 13.8. (a) The fence of the Goats-R-Us farm is triangular in shape. Two goats are tied to the fence with ropes, each at the midpoint of two different sides. The lengths of the ropes are equal to half the lengths of

---

[1] Most of the goat problem statements are taken from the delightful article "Goat on a tether" by V. Krupsky and A. Orlov in *Kvant*, 1974, issue 5.

the section of fence they are tied to. Can the goats eat all of the grass inside the fence? (b) What if the fence is a quadrilateral, and there are goats tied at the midpoints of each of the four sides with ropes that are half the length of the side they are tied to?

Additional Problems

Problem 13.9. "Peter was 10 the day before yesterday. Next year, he will be 13." Is this possible?

Problem 13.10. All of the trees in a forest are taller than 5 m and shorter than 30 m and the distance between any two trees is no more than the difference in their heights. Prove that it is possible to start from any one tree and walk around to all of the other trees and return to the initial tree having covered no more than 50 m.

Problem 13.11. This problem is adapted from Martin Gardner's *Aha! Insight*, Freeman, 1978 (reprinted in *"Aha! A two-volume collection*, MAA, 2006, p. 121).

Once upon a time, there was a tribe for which the hippopotamus was sacred. The Chief of the tribe cared for a pet hippo which he fed and coddled incessantly. Once a year, the Chief loaded his hippo and his Collector of Offerings on a boat, and headed up the river to a hut where all of his people must bring their annual offering. The hut contained a huge beam balance. The hippo was placed on one side, and gold ingots offered by the tribesmen were

placed on the opposite side until the sides balanced. One year, the hippo had grown so fat that he broke the beam balance. The Chief grew incensed and demanded that his Collector of Offerings figure out a way to measure the year's offerings. "If I do not have my offerings by sundown, your head will be mine!" cried the Chief. The Collector of Offerings thought and thought, and by the end of the day had a solution. Can you figure out what he thought of?

Winter Competition

Problem W.1. A mailman takes mail out of a public mailbox five times a day. If he opens the mailbox at equal time intervals starting at 7 a.m. and ending at 7 p.m., what is the length of each time interval?

Problem W.2. All the angles of a given pentagon are equal. Is it necessarily a regular pentagon?

Problem W.3. If only two pieces of bread fit in the frying pan at one time and it takes one minute to fry one side of each piece, what is the least time needed to fry both sides of three pieces of bread?

Problem W.4. Four balls, each of which is either black or white, lie out of sight in a box. For each of 100 tries, a person takes two balls out of the box, looks at them, and puts them back in the box, after which the box is shaken to scramble the balls. In exactly 50 out of the 100 attempts, both the balls taken out were black. How many black and how many white balls are most likely to be in the box, and why?

Problem W.5. Alice spends one-fourth of her time in school, one-fifth playing volleyball, one-sixth playing video games, one-seventh on her math homework, and one-third on everything else. Can she live like this?

Problem W.6. A small rectangle entirely contained in a bigger rectangle is removed from it, leaving a gray region. Using only a pencil and a straightedge, draw a straight line that divides the gray region into two parts having equal areas.

Problem W.7. Suppose that 1000 integers are written one after another on a line. Prove that either one of these numbers is divisible by 1000, or there are several consecutive numbers whose sum is divisible by 1000.

Problem W.8. Find the sum of the angles of a five-pointed star (see diagram).

Problem W.9. A math circle leader assigned 20 problems as homework. At the next meeting, he found out that each student solved exactly 2 problems, and that each problem was solved by exactly 2 students. (a) How many students are in the circle? (b) Is it possible to set up a discussion of the problems so that each student will be able to explain one of the problems she solved and every one of the 20 problems will be explained?

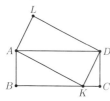

Problem W.10. (a) Show that the rectangles $ABCD$ and $AKDL$ on the left have the same area. (b) Show that the rectangles $ABCD$ and $AKLM$ on the right have the same area.

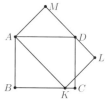

Problem W.11. The intersection of two equilateral triangles is a hexagon having three pairs of parallel opposite sides, as shown on the right. Find the perimeter of the hexagon if the perimeters of the two triangles are 9 cm and 12 cm.

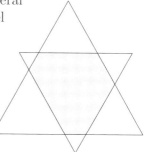

Problem W.12. Six digits are written on the board. Is it possible to arrange them so that the difference between the sum of the first three and the sum of the last three is less than 10?

Problem W.13. Suppose integers a, b and c are such that $ax^2 + bx + c$ is divisible by 5 for any integer x. Prove that each of a, b and c is divisible by 5.

Problem W.14. Find the last four digits of the number 5^{1000}.

Problem W.15. Given 20 integers, none of which is divisible by 5, prove that the sum of the 20th powers of those 20 integers is divisible by 5.

Problem W.16. The diagram shows a school, S, and Carol's house, H. If she must cross a street perpendicular to it, what is the shortest path from her house to the school?

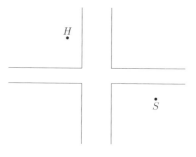

Problem W.17. A traveler starts at point A, walks 1 mile to the north, 1 mile to the east, 1 mile to the south, and ends up at point A again. Where on the surface of the earth can this occur? Find all answers.

Problem W.18. Two mirrors are attached end-to-end forming an acute angle. A ray of light is shined on one of the mirrors and is reflected according to the usual law that the angle of incidence is equal and opposite to the angle of reflection. Prove that the ray will make only a finite number of reflections in the mirrors.

Problem W.19. At dawn, two tourists simultaneously left points A and B along he same path with each headed toward the other point. They passed each other at noon without stopping. The first came to point B at 4 p.m., and the second came to point A at 9 p.m. If each walked at a constant rate, at what time did the sun rise that day?

Problem W.20. Suppose that five lattice points on a piece of grid paper are chosen, and every pair of these points is connected by a line segment. Prove that the midpoint of at least one of these line segments is a lattice point.

Problem W.21. Seventy-seven glasses are put upside-down on a table. During each round, a player is allowed to flip any four glasses over. Is it possible to flip all the glasses right-side up?

Problem W.22. (a) Five identical balls are moving in the same direction and are equally spaced along a straight track, while five more identical balls are moving in the track towards them in the same manner. All the balls are moving at the same speed, and, when they collide, the balls rebound in opposite directions. How many collisions between balls will there be? (b) What if the distances between adjacent balls can be different at the start? (This is harder.)

Problem W.23. (a) A triangle is drawn on a grid with all vertices at lattice points. Prove that the area of the triangle is equal to either the area of a whole number of grid squares, or it differs by half the area of a grid square from the area of a whole number of grid squares. (b) Prove that it is impossible to construct an equilateral triangle on a grid so that all the vertices will be at lattice points. The fact that $\sqrt{3}$ is an irrational number might prove helpful.

Problem W.24. Eighteen 2×1 dominoes cover a 6×6 board without overlapping each other or the sides of the board. Prove that, for any such arrangement, it is possible to cut the board into two pieces along a vertical or horizontal line without cutting a single domino.

Problem W.25. Chanelle and Erin alternate turns in a game. During one turn, Chanelle can put an X in any two free squares on an infinite sheet of grid paper. During one turn, Erin can put a 0 in any free square. Chanelle wants to put 100 adjacent X's in a row. Can Erin stop her from doing this?

Problem Set 14

Problem 14.1. If the number a is greater than the number b, arrange the numbers a, b and $(a+b)/2$ in ascending order.

Problem 14.2. (a) There are three apples on a table. The first weights 200 g, the second 300 g, and the third 400 g. Gabriela and Sam each take an apple and start to eat at equal rates. Whoever finishes an apple first takes the last one. If each one wants to eat as much as possible, which apple should Gabriela take first? (b) What if there is a fourth 450 g apple on the table?

Problem 14.3. Is it possible to color eight points on a line blue so that any blue point is the midpoint of a line segment with blue endpoints?

Problem 14.4. (a) Numbers have been written at the vertices of a triangle so that each is equal to the average of its two neighbors. Prove that all three numbers are the same. (b) Solve the same problem with a decagon.

Problem 14.5. In the diagram we see that a square has been cut into 4 parts, and a rectangle has been assembled from these parts. As a result, we have obtained $5 \times 13 = 65$ squares by cutting and rearranging $8 \times 8 = 64$ squares. Where is the error?

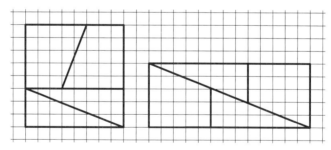

Problem 14.6. Placing weights on one side of a beam balance and putting grain on the other side until they balance will constitute one weighing of grain. We are interested in comparing sets of weights to the different amounts of grain that can be measured in one weighing. (a) One can use five 1 kg weights, and, with one weighing each, measure grain in amounts of 1, 2, 3, 4, and 5 kg. This is an example of five standard weights being able to measure exactly five different weights of grain. Choose five weights that would allow you to measure more than five different weights of grain.

How many different weights of grain result from your choice of weights? (b) What is the smallest number of weights needed to measure, in one weighing, any whole number of kilograms of grain from 1 to 7? (c) Answer the same question as in (b), but this time the range is from 1 to 77.

Problem 14.7. Queens have been placed on a chessboard so that each can capture exactly k others. (a) Find an example of such an arrangement for $k = 1$, $k = 2$, $k = 3$. (b) Can there be examples with $k = 4$? (c) Can k be larger than 4? (Queens move any number of squares across, down, or diagonally.)

Problem 14.8. Two towns are separated by two rivers, as in the diagram. The banks of each river are parallel lines. Where should bridges that cross the rivers perpendicular to the banks be built so that the distance from one town to the other is minimized?

Additional Problems

Problem 14.9. Suppose there are seven natural numbers such that the sum of any six is divisible by 5. Is each number necessarily divisible by 5?

Problem 14.10. Citizens of Polyglotia can converse in 2000 languages. Each language is spoken by more than half the citizens. Prove that it is possible to choose 10 citizens who will collectively know all 2000 languages. In other words, if every person among the 10 chosen writes a list of the languages they know, these 10 lists will together include all 2000 languages.

Problem 14.11. Find the sum of the angles $MAN+MBN+MCN+MDN$ in the diagram.

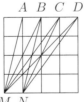

Problem Set 15

Problem 15.1. A clever salesman is counting out envelopes for a buyer. Every package contains 100 envelopes. The salesman can count out 10 envelopes in 10 seconds. How many seconds will be needed to count out 60 envelopes?

Problem 15.2. (a) Prove that a number is divisible by 4 if and only if its two final digits form a number that is itself divisible by 4. (b) Find and prove a similar rule for 8.

Problem 15.3. Jennifer thinks of a number between 1 and 32. Sandy can ask Jennifer any questions she wants to, as long as the answer can be "yes" or "no". (a) How can Sandy figure out Jennifer's number with at most five questions? (b) Can Sandy always figure out Jennifer's number by asking at most four questions?

Problem 15.4. Pick a number from 1 to 15 and find all occurrences of it in the table on the right. Tell the teacher, by row number, which rows your number is in, and the teacher will immediately guess your number without looking at the table. Can you do the same without memorizing the table?

1	7	3	15	13	11	5	9
2	15	11	6	10	14	7	3
4	5	15	6	12	7	13	14
8	12	10	14	9	13	15	11

Problem 15.5. Helen and Gary simultaneously jump from a raft on a river and swim in opposite directions. Gary swims downstream with the current at one rate and Helen swims upstream against the current at possibly a different rate. After 5 minutes they turn around and return to the raft with each maintaining their constant rate of swimming throughout. Who returns first?

Problem 15.6. (a) Find the remainder on dividing the numbers 10, 100, 1000, 10000, ... by 9. (b) Let a be a digit. Prove that the remainder on dividing the number $a0\ldots 0$ by 9 is equal to a if $a \neq 9$, and is 0 if $a = 9$.

(c) Prove that an integer is divisible by 9 if and only if the sum of its digits is also divisible by 9.

Problem 15.7. Suppose that angle A of a triangle ABC is greater than angle B. Prove that $BC > AC$.

Problem 15.8. Suppose there are exactly 9 towns in a small country, and all distances between towns are different. A person starts in each town and walks towards the closest town. Prove that: (a) There exist two towns A and B such that a person from A walks to B, and a person from B walks to A; (b) There exists a town that nobody walks to.

Additional Problems

Problem 15.9. The sum of the digits of $100! = 1 \times 2 \times 3 \times \cdots \times 99 \times 100$ has been written down in decimal notation. The sum of the digits of the resulting number is written down, and so on. Eventually the result is a one-digit number. Find that number.

Problem 15.10. A certain country has several airfields. The distances between all of them are different. An airplane takes off from each of the airfields and flies to the closest airfield. Prove that at most 5 airplanes will land at each airfield.

Problem 15.11. Vivian is thinking of a number between 1 and 16. Angela can ask Vivian any "yes" or "no" question. Vivian is allowed to lie in at most one of her answers, but Angela does not know which answer Vivian will choose for this. If Angela is allowed seven questions, how can she find Vivian's number?

Problem Set 16

Problem 16.1. Put digits in place of the asterisks so that the number $32*35717*$ is divisible by 72.

Problem 16.2. Pete and Jack are riding down an escalator. Halfway down the escalator, Jack grabbed Pete's hat and threw it onto the up escalator. Pete ran up the down escalator in order to run down the up escalator after his hat. Jack ran down the down escalator in order to run up the up escalator after the hat. Who will be first to the hat? Assume the speeds of the boys relative to the escalator are equal and do not depend on the direction of motion.

Problem 16.3. The sum of the digits of an integer is divisible by 27. Is the number necessarily divisible by 27?

Problem 16.4. Baron Münchhausen once built a fence around his lands and marked it on a map. The fence is long gone, and now the Baron cannot remember whether or not the village of Hausenhoff is part of his possessions. Fortunately, he has been able to find a fragment of the map containing his castle and the village (see figure). He knows that the fence was shown on the map as a closed dashed curve without self-intersections. Is the village on the baron's land?

Problem 16.5. Each square of a 9×9 board has a bug sitting on it. On a signal, each bug crawls onto one of the squares which shares a side with the one the bug was on. (a) Prove that one of the squares is now empty. (b) Can the bugs move so there will be exactly one empty square?

Problem 16.6. Point O is inside triangle ABC. Prove that $AO + OC < AB + BC$.

Problem 16.7. In a certain country all distances between towns are different. Jose leaves town A and heads to town B, which is the farthest town from A. From B he heads to town C, which is the farthest from B, and so on. It so happens that A and C are different towns. Prove that if Jose continues in the same way, he will never come back to A.

Additional Problems

Problem 16.8. (a) A point has been chosen on a plane. Arrange several circles in the plane so that they do not touch either the point or each other, but "hide" the point in that any ray emanating from the point meets one of the circles. (b) What is the least number of circles needed for this?

Problem 16.9. James has chosen a point within an equilateral triangle of side 1. He finds the sum of the distances from that point to the sides of the triangle. Can you tell what James' number is with certainty?

Problem 16.10. Several roads leave a highway and head to small towns, as shown in the diagram. Where on the highway should a bus stop be placed so that the sum of the distances, via roads and the highway, from the stop to each of the towns is minimized?

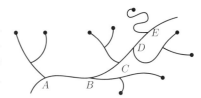

Problem Set 17

Problem 17.1. Cut the L-shaped figure in the diagram into four similar figures each half the size of the original.

Problem 17.2. Cowboy Joe walked into a saloon. He ordered a bottle of whiskey priced at $3, a pipe priced at $6, three packets of tobacco, and nine boxes of matches whose prices he did not know. The barkeep asked for $11.80. Joe drew his revolver and the barkeep recalculated and found his error. How did Joe know the first total was wrong?

Problem 17.3. A new chess piece called a camel can move in a $(1, 3)$ pattern; that is, it moves to any adjacent square and then three squares in a perpendicular direction, as shown on the diagram. Can the camel get from its original position to a square adjacent to it? Two squares are adjacent if they share a side.

Problem 17.4. Each of the diagrams below shows a lake. The points are swimmers and the circles are waves generated by the swimmers. Find the swimmers' speeds and approximate directions if the speed of the waves is $0.5\,\text{m/s}$.

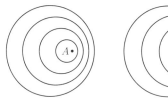

Problem 17.5. The midpoints of two sides of a triangle have been marked. How can the midpoint of the third side be found using only a pencil and a straightedge?

Problem 17.6. A castle is laid out in the shape of a 7×9 rectangle drawn on grid squares. Each square is a room, and there is a door between any two adjacent rooms. Is it possible to start in some room, walk through all of the other rooms and return to the starting place without walking through any room twice or leaving the castle?

Set 17   39

Problem 17.7. Two palm trees grow on opposite sides of a river. One is 10 m tall and the other is 15 m. The distance between the bases of the trees is 25 m. In the crown of each tree sits a bird. A fish surfaces in the river between the trees, and the two birds dive toward it simultaneously. If the birds fly along straight lines at equal rates and reach the fish at  the same time, how far from the base of the shorter palm did the fish surface? (The artist drew the trees bent, but for the purposes of the problem you should assume they are vertical.)

Additional Problems

Problem 17.8. On each square of a 9×9 grid there is a bug. At a signal, each bug crawls along a diagonal to a square sharing a vertex with the bug's square. After this move, some squares end up holding more than one bug and some squares are empty. Find the least possible number of empty squares.

Problem 17.9. Cross out one of the factors of the form $n!$ in the product $(1!)(2!)(3!)\ldots(99!)(100!)$ so that the result is the square of an integer. (Recall that $1! = 1, 2! = 1 \cdot 2, 3! = 1 \cdot 2 \cdot 3$, and so on.)

Problem 17.10. A square has been cut into five parts, as shown in the diagram. The areas of the four outer rectangles are all equal. (a) Are the outer rectangles necessarily congruent? (b) Is the central rectangle necessarily a square?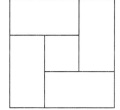

Problem Set 18

Problem 18.1. Eleven gears are connected in a chain, as shown in the diagram. Can the gears spin simultaneously?

Problem 18.2. (a) How does the area of a rectangle change if one of its sides increases by 10% while the other decreases by 10%? (b) Same question for the perimeter.

Problem 18.3. Two towns are located on a river, 10 miles apart along the water. Will it take a ship longer to go from one town to another and back, or to cover 20 miles on a lake?

Problem 18.4. The numbers $1, 2, \ldots, 10$ are written on the board. A person can pick two numbers, a and b on the board, erase them, and replace them with the number $a+b-1$. (a) How many times can one repeat this process until there is just one number on the board? (b) Can you say ahead of time what this number will be?

Problem 18.5. One liter of water is contained in each of two vessels. Half of the water from the first vessel is transferred to the second vessel, then one third of the resulting water in the second vessel is transferred to the first, then one fourth of the first is transferred to the second, and so on. How much water will remain in each vessel after 100 transfers?

Problem 18.6. Paul and Jonathan take turns breaking a 6×8 chocolate bar (it has troughs that divide it into many smaller rectangles). The one whose turn it is chooses a piece and breaks it along a trough into two pieces. The winner is the one who makes the last break. If Paul goes first, who will win?

Problem 18.7. A mouse is gnawing a cube of cheese that has been cut into 27 unit cubes. When the mouse finishes any one unit cube, it moves to a cube that shares a face with the cube it just finished. Can the mouse finish the entire cube except for: (a) a corner unit cube; (b) the central unit cube?

Problem 18.8. The midpoints of all three sides of a triangle have been marked. The triangle is then erased, leaving only the marked points. How can the triangle be recreated using only a compass and a straightedge?

Problem 18.9. A snail is oozing from point A at a constant speed, and changes direction by 90 degrees every 15 minutes. Prove that the snail can return to point A only after a whole number of hours.

Problem 18.10. A straight road crosses a field. A bus moves along the road at 10 mph. Find all of the points in the field from which it is possible to catch the bus by running at the same speed as the bus.

Additional Problems

Problem 18.11. Suppose k nickels have been laid touching each other to form a chain. One more nickel is rolled along their outer edges without slipping, but touching each nickel in the chain in turn. How many revolutions about its center will the rolling nickel make in returning to its original position if: (a) $k = 1$; (b) $k = 2$; (c) $k = 3$ and each fixed nickel is touching the other two?

Problem 18.12. Each of three identical jars is $\frac{2}{3}$ full of paint of different colors. Any part of the paint in one jar can be poured into another jar, and the paints mix homogeneously when this happens. How can the same mixture be obtained in all three jars if paint cannot be poured either out or into any other container?

Problem Set 19

Problem 19.1. Five pieces of a chain have been brought to a blacksmith. Each piece is three links long. What is the smallest number of links the blacksmith must open and reattach in order to form one chain from the five pieces?

Problem 19.2. (a) Can the asterisks in $1*2*3*4*5*6*7*8 = 0$ be changed to plus or minus signs so that the statement is true? (b) Same question for $1*2*3*4*5*6*7*8*9 = 0$.

Problem 19.3. The center of a circle circumscribed about a triangle is the midpoint of one of the triangle's sides. Prove that the triangle is a right triangle.

Problem 19.4. Is it possible to place 25 numbers in a 5×5 table so that the sum of the numbers in any column is positive, and the sum of the numbers in any row is negative?

Problem 19.5. Jamie has drawn five pictures. Each picture consists of several straight lines with the points of their intersection marked. The first picture has one point marked, the second two, the third three, the fourth four, and the fifth five. (a) Provide examples of such pictures. (b) For which numbers of marked points can we say for sure how many lines Jamie drew?

Problem 19.6. (a) How many times on May 17 do the minute and hour hands on a 12-hour clock coincide? (Count midnight only once.) (b) How many times do they form a 90-degree angle?

Problem 19.7. A boat goes downriver from town A to town B in three days, and goes upriver from B to A in five. How long will it take a raft to float from A to B?

Problem 19.8. A straight road crosses a field. A bus moves along the road at 10 mph. Find all of the points in the field from which it is possible to catch the bus by running at a speed of 5 mph.

Problem 19.9. A number ending in 01 has been written on the board. Prove that a line can be drawn between two adjacent digits of this number so that the number of 1's to the left of the line equals the number of 0's to the right of the line.

Problem 19.10. An ant is sitting at the midpoint of one of the sides of an equilateral triangle. The ant wants to visit each of the triangle's remaining sides and return to the original point. How can it do this in a way that minimizes the distance it travels?

Additional Problems

Problem 19.11. Suppose we have points M and N, and a circle with diameter AB positioned in a plane as in the diagram. How can you use only a pencil and a straightedge to drop a perpendicular from M onto AB and from N onto the continuation of AB?

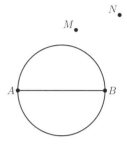

Problem 19.12. Can a 10×10 square be paved with 1×4 rectangular stone plates?

Problem 19.13. A marble has been placed in the square at the far right end of a 1×20 grid. Two players take turns moving the marble right or left a number of squares as long as that number has not been used in any prior turn. The winner is the last to make a move. Which of the players, the first or second to move, can always win, and how should he play to do so?

Problem Set 20

Problem 20.1. Two clocks started to tick at the same time. The first ticks every 2 seconds; the second every 3 seconds. Ticks that occur simultaneously are heard as one tick. How much time will pass between the first and 13th ticks?

Problem 20.2. Suppose we have two cups. One has 150 ml of milk, the other has 150 ml of coffee. A teaspoon of milk is taken from the first cup and poured into the second cup. The liquid is mixed, and then a teaspoon of the mixture is taken from the second cup and poured into the first cup. Is there more coffee in the first cup or more milk in the second cup?

Problem 20.3. Base AD in trapezoid $ABCD$ is longer than base BC. Which is greater, the sum of angles A and D, or the sum of angles B and C?

Problem 20.4. A notepad costs a whole number of cents. Nine notepads cost less than \$10, and ten notepads cost more than \$11. How much does one notepad cost?

Problem 20.5. The height dropped from A onto BC in triangle ABC is not shorter than BC, and the height dropped onto AC from B is not shorter than AC. Find the angles of ABC.

Problem 20.6. A line of recruits stands facing their drill sergeant. When he gives the order "left face!", several turn to the left, several to the right, and the rest turn around and face the opposite way. Can the sergeant find a place in the formation so that the number of recruits to either side who are facing him will be equal?

Problem 20.7. The numbers $1, 2, \ldots, 10$ are written on the board. Any two numbers a and b can be erased and replaced with the number $a - b$. After several such turns, can the only number remaining on the board be 0?

Problem 20.8. What is the smallest possible number of students in the math circle if the girls in the circle are less than 50% but more than 48%?

Problem 20.9. A straight road crosses a field. A man, standing on the road at point A, can walk in the field with a speed at most 3 km/h and along the road with a speed at most 6 km/h. Draw all the points the man can reach in 1 hour.

Additional Problems

Problem 20.10. Two players play a game on a 10×10 white board. At each of his turns, the first player can paint in black any 4 squares forming a 2×2 square. At each of her turns, the second player can paint in black a pattern of 3 squares forming a corner (that is, a 2×2 square minus a 1×1 square). The two players alternate turns. Already painted squares cannot be repainted. The winner is the last to be able to paint squares. Which of the two players can always win, regardless of the moves of the other, and what strategy should be used to guarantee a win?

Problem 20.11. A $4 \times 4 \times 4$ cube has to be cut into 64 unit cubes. What is the least number of cuts required to do so if one can rearrange pieces after each cut? Every cut must be parallel to a face of the big cube.

Problem Set 21

Problem 21.1. Find the smallest integer greater than 40500 which is the square of an integer.

Problem 21.2. Prove that (a) $a^2+b^2 \geq 2ab$ for any a and b; (b) $x+1/x \geq 2$, if $x > 0$; (c) $a^2/b \geq 2a - b$ if $b > 0$.

Problem 21.3. (a) Is it possible to write eleven numbers in a row so that the sum of any three consecutive numbers is positive, while the sum of all the numbers is negative? (b) Same question with twelve numbers.

Problem 21.4. (a) Prove that any rectangle with a perimeter of $4\,\text{cm}$ has an area of at most $1\,\text{cm}^2$. (b) Is there a rectangle with a perimeter of $4\,\text{cm}$ whose area is exactly $1\,\text{cm}^2$?

Problem 21.5. In a triangle ABC, a line parallel to AC has been drawn through the point of intersection of the angle bisectors of the triangle. That line intersects AB and BC in points M and N, respectively. Prove that $MN = AM + CN$.

Problem 21.6. A cashier counts money as follows: first he counts how many bills he has regardless of denomination, then he adds the number of bills with denomination greater than \$1, then the number of bills with denomination greater than \$2, and so on. Why will he end up with the correct dollar total?

Problem 21.7. Prove that for any numbers $x, y,$ and z, we have
$$x^2 + y^2 + z^2 \geq xy + yz + zx.$$

Problem 21.8. (a) Blindfolded, you enter a room where there are two tables. The table to your left has a layer of nickels one nickel thick, and the one to your right is empty. You're wearing gloves, so you can't tell which coins are heads up, but you know that 31 coins are tails up and the rest are heads up. You can transfer any number of coins from either table to the other and, at the same time, flip over any desired number of coins. Can you arrange things so that, possibly after several transfers, the number of nickels that are tails up is the same on both tables? (b) What if the initial situation is that there are nickels on both tables, with 31 of those on the left table and 8 of those on the right table tails up?

Set 21   47

Problem 21.9. Two perpendicular straight roads cross a field. A person standing at their intersection can walk across the field no faster than 3 km/h, and along the roads no faster than (a) 6 km/h; (b) $3\sqrt{2}$ km/h. For each case, draw all of the points the person can reach in 1 hour.

Additional Problems

Problem 21.10. A farmer, who has come to market with walnuts to sell, finds himself with an accurate 1 kg weight and an inaccurate beam balance whose arms are not equal. To weigh out 2 kg of walnuts for a customer, the farmer first puts the weight in the right pan of the balance and adds nuts in the left until they balance. Next he balances the weight in the left pan with nuts in the right. (a) Prove the resulting bag of nuts weighs more than 2 kg.  (b) How can the farmer weigh out exactly 2 kg of nuts?

Problem 21.11. The numbers from 1 to 64 have been placed in each square on a chess board so that each number appears exactly once. Prove that there will be two squares sharing an edge whose numbers will differ by more than 4.

Problem Set 22

Problem 22.1. Anna, Bryan, Caroline and Dustin have been collecting mushrooms. Anna collected more mushrooms than anyone else; Caroline did not collect the least. Is it true that the girls collected more mushrooms than the boys?

Problem 22.2. Suppose that n is a natural number. Can the number $n(n+1)$ be a perfect square?

Problem 22.3. A plane has been colored with two colors, and both colors have been used. Prove that two points 1 inch apart can be found that are colored (a) the same color; (b) different colors.

Problem 22.4. Can numbers be arranged in a table so that the sum of 4 numbers in each 2×2 square is negative, while the sum of all the numbers in the table is positive if the table is (a) 6×6; (b) 5×5?

Problem 22.5. Find all natural numbers m and n so that $m^2 = 57 + n^2$.

Problem 22.6. A car's front tires need to be replaced after 25000 miles, and the rear ones need replacing after 15000 miles. When should the tires be rotated, front to back, so that they need to be replaced simultaneously? How far will the car travel in total? Assume that the tires wear out at constant rates.

Problem 22.7. The product of two positive numbers is greater than their sum. Prove that their sum is greater than 4.

Problem 22.8. There are 15 left boots and 15 right boots in a row. Prove that it is possible to choose 10 consecutive boots for which the numbers of left and right boots are equal.

Additional Problems

Problem 22.9. (a) Five congruent paper triangles have been placed on a table. Each may be shifted in any direction but not rotated or flipped over. Is it always possible to cover each one of these triangles with the other four? (b) Answer the same question, but suppose the triangles are equilateral in addition to being congruent. (c) Under the conditions in part (b), is it always possible to cover any of the triangles with just three of the remaining ones?

Problem 22.10. A circle has been colored with two colors. Prove that there will be three points of the same color that form an isosceles triangle.

Problem 22.11. Suppose that the numbers a, b, c, and d are positive. Prove that
$$\frac{a^2}{b} + \frac{b^2}{c} + \frac{c^2}{d} + \frac{d^2}{a} \geq a + b + c + d.$$

Problem Set 23

Problem 23.1. Think of two 2-digit numbers. Multiply your first number by 4; add 7 to the product; multiply that result by 25; add in your second number, and finally add 125. Tell the instructor your result and he will tell you your original two numbers. How does he know them?

Problem 23.2. Is it possible to cut a triangle into two acute triangles with one straight line cut?

Problem 23.3. John multiplied two natural numbers that differed by 1, while Coleen multiplied two natural numbers that differed by 2. Could they end up with the same number?

Problem 23.4. Suppose that x is a natural number, and suppose that two of the following five statements are false and three are true: $2x > 70$; $x < 100$; $3x > 25$; $x \geq 10$; and $x > 5$. Find x.

Problem 23.5. Find a point on the hypotenuse AB of a right triangle ABC for which the sum of the squares of the distances to AC and BC is the least.

Problem 23.6. Some blue ink has been spilled onto a white plane. (a) Prove that you can always find an equilateral triangle with vertices of one color. (b) Is it always possible to find such a triangle with a side of 1 foot?

Problem 23.7. Find all natural numbers x and y for which $1/x + 1/y = 1/2$.

Problem 23.8. A hermit crab left a beach at 9 a.m. one morning and began to climb up a cliff. It reached the top of the cliff at 8 p.m. At 9 a.m. the next morning, the crab began its descent, using the same path. It reached the beach at 8 p.m. Prove that there is a point on the cliff that the crab passed at the same time going up and coming down. Do not assume the crab's speed is constant.

Set 23   51

Additional Problems

Problem 23.9. Find all natural numbers a, b, c such that $a! + b! = c!$.

Problem 23.10. A sail, as shown in the diagram, has the form of the quadrilateral $ABCD$. The angles A, B, D measure $45°$ each, while C measures $225°$. If $AC = 4$ meters, find the area of the sail.

Problem Set 24

Problem 24.1. A square wheel is rolling down a straight road. Draw the path of the center of the wheel.

Problem 24.2. Two volumes of Edgar Allan Poe's works are adjacent on a shelf. In each volume, the pages are 2 cm thick, and each front and back cover adds 2mm. A bookworm chewed a path perpendicular to the pages from the first page of the first volume to the last page of the second volume. What distance has the bookworm chewed through?

Problem 24.3. There are 28 students in a class. Every girl has four friends among the boys in the class, while every boy has three friends among the girls in the class. How many boys and how many girls are there in the class?

Problem 24.4. A clock shows 12:00. How long will it be before the hour and minute hands coincide again?

Problem 24.5. A line is colored with two colors, red and green. Prove that it is possible to find a segment of the line for which the endpoints and midpoint are of the same color.

Problem 24.6. An airport has fifteen gates and various moving walkways, each of which connects exactly two gates in both directions. From each gate you can reach at least seven others via a single walkway. Is it possible to go from any gate to any other via either one or two walkways?

Problem 24.7. Find all natural numbers x and y for which $x + y = xy$.

Problem 24.8. A right triangle has area S, and its sides form diameters of semicircles, as in the diagram above. Find the joint area of the crescents these semicircles form, shown with red stripes. You can use the fact that the area of a circle of radius r is πr^2, with $\pi = 3.1415926\ldots$.

Problem 24.9. Baron Münchhausen claims he once drew a polygon having a point in its interior with the property that no edge of the polygon could be seen in its entirety from that point. Is he telling tall tales again?

Set 24   53

Additional Problems

Problem 24.10. An equilateral triangle AKB has been constructed outside of the square $ABCD$ on the side AB. Find the radius of the circle circumscribed about triangle CKD if $AB = 1$.

Problem 24.11. Color a plane with three colors so that every line is colored with exactly two colors.

Problem 24.12. Fifteen rooks have been placed on a 15×15 board so none can capture any other. Each now moves in a knight's move. Prove that at least two rooks can now capture each other.

Problem Set 25

Problem 25.1. A man was born on May 1, 30 BCE, and died on May 1, 30 CE. How many years did he live?

Problem 25.2. (a) Terry walks on a trail and meets a group of five people who walk in the opposite direction. Prove that in this group, either at least three people know Terry, or at least three people don't know her. (b) Prove that among any six people, it is always possible to find three that know one another, or else there are three mutual strangers.

Problem 25.3. The average age of 11 soccer players on a team is 22. During a game, one player got injured and left the field. The average age of the remaining 10 players became 21 years. How old is the injured player?

Problem 25.4. An apartment is a 3×3 square divided by the walls into square rooms, each of which is 1×1. There is a door in the wall between any two adjacent rooms, and currently all these doors are closed. There is a cat inside one of the rooms. Find the least number of doors that must be opened to let the cat wander through the entire apartment.

Problem 25.5. A group of islands is connected by bridges in such a way that it is possible to get from each island to any other one. A tourist has visited all the islands and, while doing so, he walked across each bridge exactly once. He visited Clearview Island exactly three times. How many bridges emanate from Clearview Island if the tourist (a) did not start at Clearview Island and did not end there; (b) started at the island but ended at a different one; (c) started and ended at Clearview Island?

Problem 25.6. Numbers are placed into the cells of a rectangular table in such a way that the sum of the numbers in each column and in each row is 19. Prove that the table must be a square.

Problem 25.7. A bug is crawling along the edges of a cube. Will it be able to move along all the edges of the cube while crawling along each edge exactly once?

Set 25   55

Problem 25.8. Is it possible to draw a polygon and a point outside of it so that none of the edges of the polygon is completely visible from the point?

Problem 25.9. A number of students are standing in a circle. The teacher will select one to go to the cafeteria and get cookies for the whole class by the following process. One student is picked and is told to remain standing. The next student, going counterclockwise, is told to leave the circle and sit down. The next student is told to remain, and the next one is told to go sit down. The teacher continues in this way, telling every other student (always advancing counterclockwise) to go sit down. The circle gets smaller and smaller until there is only one student left standing. What place did the last student originally occupy if the number of students was (a) 16; (b) 17; (c) 34; (d) 1000 to begin with?

Additional Problems

Problem 25.10. Ivan, Victor, and Ashley play table tennis. Two players play at a time, and the third person plays the winner of the previous game. There are no ties. Eventually, Ivan played 10 games while Victor played 21 games. How many games has Ashley played?

Problem 25.11. A number of wallets hold 1000 dollars in all, and the wallets were placed in a number of pockets. If no wallet contains another wallet, and the total number of wallets is greater that the number of dollars in any pocket, is the number of pockets greater than the number of dollars in some wallet?

Problem Set 26

Problem 26.1. Two carts were moving towards each other, one at the speed of 10 km/h, and the other at 15 km/h. When the distance between the carts was 50 km, a fly left the first cart and flew towards the second one. When it reached the second cart it turned around and flew towards the first cart; when it reached it, the fly turned around again and flew towards the second cart, and so on. If the fly's speed is 20 km/h, how many kilometers will the fly have flown by the time the carts meet?

Problem 26.2. What is the least number of pieces a 12 cm long wire can be divided into to make a cube whose sides are 1 cm? Pieces of the wire can be bent and connected together.

Problem 26.3. (a) A square hole has been cut out along grid lines from a square piece of grid paper. Can the remaining paper have exactly nine grid squares? (b) How about ten grid squares?

Problem 26.4. A certain country has exactly $N > 1$ towns, as well as some roads connecting some of the towns. Suppose there is a unique way to get from each town to every other one by traveling along the roads and never going on any of the roads more than once. (a) Prove that there is a town with exactly one road emanating from it. Also try to prove that there are two such towns. (b) How many roads are in the country? (c) One road has been closed for repairs. Is it still possible to get from each town to every other one?

Problem 26.5. A road connecting two mountain villages has no horizontal stretches. A bus goes uphill at a constant speed of 15 mph, while it goes downhill at a constant speed of 30 mph. What is the distance between the villages if a roundtrip takes four hours?

Problem 26.6. (a) Is it possible to arrange four balls in space so that every ball touches the other three? (b) Same question, except now we want five balls, each touching the other four. The balls can be of different sizes.

Problem 26.7. A volleyball net is a rectangle made up of squares arranged in a 50×600 grid. Edges of these squares are ropes, and their vertices are knots connecting the ropes. What is the largest number of ropes that can be cut without having the net come apart into two or more pieces?

Set 26   57

Problem 26.8. A fuse burns from end to end in an hour, but it does not burn uniformly. How can we measure exactly 45 minutes if we have two such fuses, each of which burns non-uniformly in its own unique way?

Additional Problems

Problem 26.9. Three circles pass through a point as shown in the picture. Find the sum of the angles of the red curvilinear triangle.

Problem 26.10. Seventeen points have been picked in a plane, and each pair of points has been connected by a line segment of one of three colors: red, yellow, or green. Prove that there are three points which are the vertices of a monochrome triangle.

Problem Set 27

Problem 27.1. Is it possible to choose three natural numbers x, y, and z so that $28x + 30y + 31z = 365$?

Problem 27.2. The fair had come to town, and Andy and his dad went to the shooting gallery. His dad has agreed to let Andy shoot five times. For each hit, Andy got two more shots. He had a total of 25 shots. How many hits did he have?

Problem 27.3. Three students are riding a train pulled by a steam engine. After the train exits from a tunnel, each student notices that the faces of the other two are smudged with soot, and each one starts laughing at the other two. Suddenly the smartest student realizes that his face is dirty too. How did he figure this out?

Problem 27.4. A number has been obtained by rearranging the digits of another number. (a) Can the sum of these two numbers equal 9999? (b) Can it equal 99999?

Problem 27.5. (a) A traveler came to an inn. He did not have any money, but he had a silver chain of seven links. The innkeeper requested one link per day paid every day without delay. He refused to accept any advance payment, but agreed to return previously received links as change when needed. What is the smallest number of links that must be cut so that the traveler can stay in the inn for seven days and pay the innkeeper as required? (b) What if the number of links and days is 23?

Problem 27.6. Two white and two black knights are placed at the corners of a 3×3 chessboard as shown in the first diagram on the right. Is it possible to move them according to the rules of chess and obtain, after several moves, the arrangement in the second diagram?

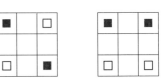

Set 27

Problem 27.7. Twenty balls are glued to form two chains of four balls each and two 2×3 rectangles as shown on the right. How can these figures be arranged to form a triangular pyramid?

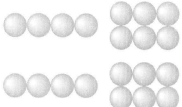
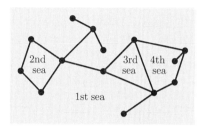

Problem 27.8. Can you provide a rational explanation at to why buckets are usually made as truncated cones rather than cylinders?

Problem 27.9. Suppose I different sea islands are connected by nonintersecting dams in such a way that it is possible to get from any island to any other one along the dams. In addition to the surrounding sea, the dams may bound little seas. Suppose that altogether there are D dams and S seas — an example of a possible arrangement is shown in the diagram. (a) Assuming that $S = 1$, show that $I - D + S = 2$. (b) Suppose that $S \geq 2$. Prove that if one dam is destroyed, then the number of seas will also be reduced by one, but it would still be possible to get from any island to any other one along the remaining dams. (c) Prove that the equation $I - D + S = 2$ always holds.

Additional Problem

Problem 27.10. One marker is placed in the center of a 9×9 board. Ann and Beth take turns moving the marker to one of the adjacent squares — one sharing a side — provided that this square has never been occupied by the marker. Ann goes first. The player unable to move loses. Which of the players can guarantee a win?

Problem Set 28

Problem 28.1. Given a water faucet and a cylindrical pot, how can you fill the pot exactly half-way with water?

Problem 28.2. Why are manhole covers circular rather than a square?

Problem 28.3. The letter P was painted in one of the squares of the assembly of seven congruent squares shown on the right. A cube with faces congruent to these squares was placed on the square with the P so that its face coincided with the square, and then the cube was rolled over its edges along the figure. In the process, the letter was imprinted on the face of the cube and also on all squares of the figure where this face landed. Which of the squares of the figure got an imprint of the letter, and what exactly do those imprints look like?

Problem 28.4. Is it possible to place six identical unsharpened pencils in a configuration where each pencil touches all the other ones?

 Problem 28.5. (a) The left half of the picture on the left shows the path of a fish as seen from the front of an aquarium. The picture next to it shows the same path as seen from the right side of the aquarium. Draw the path of the fish as seen from above. (b) Do the same problem for the paths shown on the right.

Problem 28.6. Is it true that 1-liter and 2-liter Coke bottles are proportional (one can be obtained from the other by scaling up all dimensions by the same factor)?

Problem 28.7. An unfolded cube is shown on the left half of the figure below. Which of the three cubes on the right can unfold in this way?

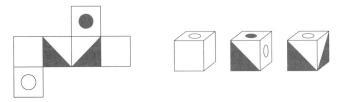

Problem 28.8. Can the shadow of an opaque cube be a regular hexagon?

Set 28

Problem 28.9. Suppose we have three identical bricks. How can we find the length of the diagonal of a brick (i.e., the distance between two vertices not belonging to the same face) using only one measurement with a ruler?

Problem 28.10. (a) Is it possible to saw a figure out of a piece of wood so that every face of the figure is a square, but the figure is not a cube? (b) What if we ask, in addition, that the figure be convex?

Problem 28.11. The picture shows that the yacht *Alpha* was moored at the dock before *Kvant*. Can *Alpha* sail first without having to take *Kvant*'s mooring cable off the bollard?

Additional Problems

Problem 28.12. In three-dimensional space, we are given a point light source emitting light in all directions. Is it possible to choose (a) 100 opaque balls, (b) 4 opaque balls, and place them in space in such a way that they don't intersect each other, don't touch the light source, and completely block its light, i.e., every light ray emanating from the bulb meets one of these balls?

Problem 28.13. You have a compass, a ruler, a piece of paper, a pencil, and a ball. You can draw on the surface of the ball using the compass and the pencil, and you can draw on the paper using the compass, the pencil, and the ruler. Can you draw a line segment on the paper whose length is the radius of the ball?

Solutions to Problem Set 0

Problem 0.1. It takes her 10 minutes. She climbs four flights of stairs to go from the first floor to the fifth floor, and climbs eight flights to go from the first floor to the ninth floor. This is exactly twice as many flights of stairs so it takes twice the time.

Problem 0.2. The fact that the pointer is shifted means the scale shifts (changes) every weight by the same quantity. Let's call this quantity x. We don't know the true weight of the first bag, but when we add x to it we get the reading 2; similarly x added to the true weight of the first bag gives the reading 3. The two readings add up to 5, which is $x + x$ different from the sum of the true weights. Weighing the bags together gives a reading of 6, which is x different from the sum of the true weights. Thus $x = -1$: the scale shows 1 lb less than the true weight. Hence the correct weights of the bags are 3 lbs and 4 lbs.

Problem 0.3. Let's watch the labels for the thumb. To begin with, it gets selected with 1, then we count four from the index finger to the pinkie and then four back again (from the ring finger to the thumb), so that the thumb gets selected on $1 + 8 = 9$. Next it gets selected on $9 + 8 = 17$, with each subsequent visit to the thumb eight more than the previous one. Since 1000 is divisible by 8, the thumb would be selected on $1 + 8 \times 125 = 1001$. Therefore, the index finger will be the 1000th one.

Problem 0.4. If the dot is at the center, we don't need to cut the circle at all. Suppose the dot is not in the center and not on the edge.

(a) Cut out two small, nonoverlapping circles of the same radius, centered at the dot and at the center of the original circle. The little circles should be completely within the given circle, as in the diagram. The original circle has been divided into three parts. Now interchange the two small circles.

(b) Solution 1: We can change the previous solution slightly. Draw two common tangents to the small circles, as shown below. Now cut out the oval-shaped figure thus obtained. We have divided the original circle into two parts; now just turn the cut-out figure through 180 degrees!

Solution 2: No matter where the dot is, draw a circle centered at the dot with the radius of the original circle, and cut along

the resulting arc that is inside the original circle. Except when the dot is at the original center, one gets a lens-shaped piece and a piece shaped like a crescent moon. Slide the crescent moon to the other side of the lens, at the same time turning it around. This yields a new circle with its center at the dot.

Problem 0.5. At the moment the brother leaves the house, his sister is a certain distance ahead of him; let's say she is x meters ahead. Then every subsequent 5 minutes the sister will walk the same distance of x meters while the brother will walk $1.5\,x$ meters. Therefore, every 5 minutes the distance between them will become $0.5x$ meters less. Since the distance between them was x meters to begin with, it would take 10 minutes for the brother to catch up.

Problem 0.6. Usually people think that if the bulldozer moved forward 10 cm then so did the point A. It's easy to see that this is wrong. If this were the case, then point A would be staying in the same place on the bulldozer, i.e., right at the middle, which would mean that the tread of the bulldozer does not move. What, then, is the distance that the point A advances?

Let's look at the point on the forward wheel of the bulldozer where it touches the ground after the bulldozer has moved. This point is 10 cm ahead of where the wheel touched the ground before the bulldozer moved. Therefore 10 cm of the tread has reached the ground as the bulldozer moved ahead 10 cm. Since point A has moved 10 cm with respect to the middle of the bulldozer, and the bulldozer has moved 10 cm forward, the point A has moved $10 + 10 = 20$ cm. This is the shift of the point A relative to the ground.

To see this a different way, let's imagine that we are by the side of the bulldozer and moving with it while recording it on a video camera. When we watch the video, we will see the bulldozer in a fixed position on the screen but with wheels rotating and the tread moving. During the screening of the video point A will move 10 cm on the screen. At the same time our camera, has gone 10 cm. Thus point A has indeed moved through 20 cm.

Problem 0.7. Since Owl and Rabbit together ate 45 bananas, one of them ate at least 23 bananas. Then Winnie the Pooh ate at least 24 bananas, and hence the three of them ate at least 69 bananas. Thus Piglet ate at most one banana. The problem states each ate at least one banana, so Piglet ate exactly one.

Problem 0.8. Label Nicole's seventh tree A and her 20th tree B. Following the direction of Nicole's count, there are 12 trees strictly between A and B. Now keep counting beyond B but use Valerie's count. There are 86 trees between B and A, since there are exactly 86 integers greater than 7 and smaller than 94. Remembering to include A and B in the final count, we see that the total number of trees is $12 + 86 + 2 = 100$.

Can you tell whether the girls walked in the same direction or in opposite directions while counting trees?

Problem 0.9. In a pair, if the boy has three times as many flowers as the girl, then this pair together has four times as many flowers as the girl. If, in a pair, the boy has one third as many flowers as the girl, then the girl has three times as many as the boy, so the total number of flowers for the pair is four times the number of flowers the boy has. In each case the total number of flowers in each pair is divisible by 4. Therefore, the total number of flowers must be divisible by 4. But 2006 is not divisible by 4 and so the whole group cannot have 2006 flowers.

Solutions to Problem Set 1

Problem 1.1. If we subtract 3.25 pounds from 5 pounds, the result is the weight of half of the water in the beaker. This is 1.75 pounds. Thus the vessel can contain twice this amount of water or 3.5 pounds.

Problem 1.2. Let's first compare the numbers $333\,333 \times 444\,444$ and $222\,222 \times 666\,666$. They are equal. The first number equals $3 \times 111\,111 \times 4 \times 111\,111$, and the second equals $2 \times 111\,111 \times 6 \times 111\,111$, but $3 \times 4 = 2 \times 6 = 12$.

Now we can solve the problem. The number $666\,667$ equals $666\,666 + 1$, so we can write

$$222\,222 \times 666\,667 = 222\,222 \times 666\,666 + 222\,222 = 333\,333 \times 444\,444 + 222\,222.$$

Thus $222\,222 \times 666\,667$ is larger by $222\,222$.

Problem 1.3. The sum of the distances will always be 1. From the point M on side AC, the distance from M to AB is the length of the perpendicular extended from M to AB. Let us denote it by MX. Triangle MXA is similar to triangle CBA, and thus $MX = AX$. But, because MX and BC are parallel lines, the distance from M to BC is the same as the distance from X to BC which is BX, as one can see in the diagram. Thus the sum is equal to $AX + XB = AB = 1$.

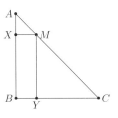

Problem 1.4. Solution 1: If we borrow one coin from Peter, he is left with between 0 and 99 coins. The remaining coins can be evenly divided by 2, 3, 4, and 5. Because 3, 4, and 5 do not have a common factor, the remaining amount of coins must be divisible by their product, which is 60. Of the integers 0 to 99, only 0 and 60 are divisible by 60, so Peter has either 1 or 61 coins. It's not very likely that Peter would start dividing one coin into piles!

Solution 2: Because the number that is one less than the amount of Peter's coins is divisible by 2, it is even, and, because it is divisible by 5, it ends with 5 or 0. It cannot end in 5, because it must be even. Thus the number is either 0, 10, 20, ..., 90. Of those numbers, only 0, 30, 60, and 90 are divisible by 3. Of those, only 0 and 60 are divisible by 4, so Peter has either 1 or 61 coins.

Problem 1.5. Suppose that the traced rectangle and the plate are positioned so their longer sides are horizontal. We can draw a 3×10 rectangle inside the traced rectangle by rotating the plate 90 degrees, alternately placing its long side on the short sides of the tracing, and drawing vertical lines. In this smaller rectangle, we can draw the diagonals using the longer side of the plate as a straightedge. The point of intersection of the diagonals will also be the center of the original traced rectangle. 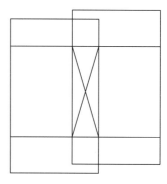 The longer side of 17 in would suffice since the diagonal of a 3×10 rectangle is shorter than 13, the sum of its sides, which in turn is smaller than 17. See the illustration.

Problem 1.6. The total number of potatoes is 12. Each eats 4, and thus Angela has given Greg 3 potatoes and Minh has given him one. The chocolate should be split the same way: Angela gets 9 pieces and Minh gets 3.

Problem 1.7. Yes, it is possible. Let us divide a square into boxes with lines parallel to its edges so that each box is $1\,\text{mm} \times 1\,\text{mm}$. Inscribe circles in each box. Each circle's diameter is $1\,\text{mm}$. There are $100 \times 100 = 10000$ such circles, and the sum of their diameters is $10000\,\text{mm}$, or $10\,\text{m}$.

Problem 1.8. It's easy to figure out if the dot is in the square with 4 lines; just draw your lines through the edges of the box. But it is also possible to do it with just 3 lines. To do this, draw the first line through one of the diagonals of the square. It divides the square into two triangular halves, and we now know in which half of the square the dot might lie. After this we draw lines through the remaining two sides of this half of the square and figure out if the dot is inside or outside the square. What could be done if the dot lies on the first line?

Why aren't two lines enough? It is because it's impossible to bound a finite region with just two lines. Regardless of how we draw those two lines, the dot might lie within the unbounded angle formed by lines that intersect the square. We have no way of determining if the dot is inside or outside the square.

Problem 1.9. (a) No, not necessarily. Let the train move at a constant speed of 90 mph for the first half hour, at 110 mph the next half hour, at 90 the next, and so on. Clearly, the train covered the first 100 miles in an hour. Now think of any one-hour time interval. During that time, the train goes 90 mph for 30 minutes of that hour and 110 mph for the remaining 30 minutes. One of these 30-minute times will most likely consist of some number of minutes at the start of the hour and the remainder of the 30 minutes at the end of the hour. The train goes 100 miles in that hour and, since the train never stops, that 100 miles is covered in exactly one hour.

(b) No. Average speed is the ratio of total distance to total time. Thus the problem can be reformulated as follows: does the train necessarily have to have moved through 550 miles over the five and a half hours? The answer is no. In moving according to the scheme outlined in part (a), the train will travel through 545 miles. Check it yourself.

(c) Some interesting questions: What is the minimum distance the train can travel if the conditions in the problem are met? What is the maximum distance?

Solutions to Problem Set 2

Problem 2.1. There is one square in the first figure, and two more in each one after that. Thus the numbers of the squares in the figures can be represented as a sequence of odd numbers: $1, 3, 5, \ldots$. In order to reach the 100th figure, it is necessary to add two squares to the first figure 99 times. Thus the 100th figure will consist of 199 squares.

Now lay down the diagrams in reverse order:

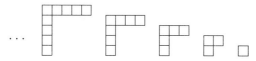

Notice that the first figure (now last) fits inside the second (now second-to-last) to form a 2×2 square. That nests inside the next figure to form a 3×3 square, and so on. Thus the first 100 figures together form a 100×100 square, which contains $100 \times 100 = 10\,000$ little squares.

Problem 2.2. (a) Of course not: the grasshopper can only jump a whole number of inches.

(b) No, the grasshopper can move only an even number of inches, so its distance from the original position will always be even.

(c) Yes; for example, it can jump 6 inches twice to the left and 8 inches once to the right.

Problem 2.3. Mark the midpoint of each side of the rectangle and connect the midpoints of opposite sides. The rectangle has now been divided into four identical smaller rectangles each of area $\frac{1}{4}$. Now divide one of these rectangles along its diagonal cutting off one-eighth of the total area, as in the diagram. The remaining area is equal to $1 - \frac{1}{8} = \frac{7}{8}$.

Problem 2.4. (a) Pick a gnome at random. That gnome has played 1 game with each of the 9 remaining gnomes, for 9 games.

(b) Solution 1: Place 10 points along the circumference of a circle. Each point represents a gnome. Connect every two points with a line segment, representing a game. We now need to count the total number of line segments. One way to do this is to count the ends of the line segments in two different ways. Erase most of the middle of each line segment leaving little "whiskers" coming out of each point. There are 10 points and each has 9 "whiskers" so there are 90 "whiskers" in all. If n is the number of line segments then there are $2n$ "whiskers", two for each segment. Therefore $2n = 90$ or $n = 45$. Thus the total number of games is 45.

Solution 2: The first gnome played 9 games; the second, 8 more, as his game with the first was already counted; the third played 7 more, as his games with the first two were counted; and so on. Thus the total number of games is equal to $9+8+7+6+5+4+3+2+1+0$. All that remains is to evaluate this sum. By the way, is there more than one way to do it? We will talk about that in subsequent problems.

Problem 2.5. When the window was opened, the center of the upper panel formed a rectangle of dimensions 10×25 inches. This rectangle is the region with two layers of glass. If we close the window, the hole disappears, and the glass is one layer all the way across the window. Thus, the glass that was in two layers when the window was open has now gone towards closing the gap. Thus the area of the opening is the same as the area of the rectangle which is 250 square inches.

Problem 2.6. Let's try to answer the following question: Can there be many nonpositive numbers among the given 2002 integers? Definitely no more than 99, for otherwise we could choose 100 nonpositive numbers whose sum would be nonpositive. Then let us choose all nonpositive numbers and enough positives so there is a total of 100 numbers. The sum of the chosen numbers will be positive by the premise of the problem, and all of the rest of the numbers will be positive. Thus the sum of all 2002 numbers will also be positive.

Problem 2.7. Let A be the corner where the ant starts and B where it ends. The room has a total of six faces (walls, floor, and ceiling). Three are adjacent to corner A, and three to corner B. To get from A to B, the ant must cross from a face adjacent to A to one adjacent to B.

Let's say this occurs on the edge XY, at point M as in the diagram. Up to M, the shortest path for the ant was along the line segment AM; after that the shortest path is the line segment MB.

Now we mentally cut out the two faces of the cube containing AM and MB and flatten them out in a plane so they form the rectangle in the diagram. For which point M is $AM + MB$ the shortest path?

The answer, of course, is when AMB is a straight line in the rectangle. Then the path is simply the diagonal AB of the rectangle. Thus, the ant should first move from A along a straight line segment to the middle of one of the edges not adjacent to either A or B, and then move from that point along a straight line to B.

Set 2

Problem 2.8. Solution 1: The speed of the motorcycle is double the speed of the truck. Draw a diagram that shows the positions of the vehicles when the bus, labeled B, passes the first observer:

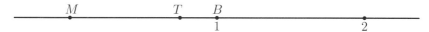

Let us say that the truck, T, was x miles from the first observer at that time. At that moment the motorcycle, M, was $4x$ miles from the first observer because, when the truck covers x miles to the first observer, the motorcycle will have covered $2x$ miles, and will still have to travel the same amount of time until it reaches the first observer, and, in doing so, it covers $2x$ more miles.

From the work above we have also determined that the intervals at which the vehicles pass the observers correspond to the time in which the motorcycle moves $2x$ miles.

Now draw the positions of the vehicles when the bus passes the second observer:

At that moment the motorcycle was at a distance of $2x$ miles from the second observer, meaning that at that moment the motorcycle and truck were at the same point because it takes twice as long for the truck to cover this distance. We have seen that this point where the motorcycle catches the truck is $2x$ beyond the first observer, so the truck has traveled $3x$ miles before they meet and the motorcycle has traveled $6x$ miles.

So the distance between the observers is $4x$ miles, and the bus covers that distance in the same time as the truck, which, at a speed of 30 mph, covers $3x$ miles. Thus the speed of the bus is 40 mph.

Solution 2: Let t be the common time interval between the moments the vehicles pass the observers. The motorcycle reaches the first observer t hours after the truck and reaches the second observer t hours before the truck. Therefore the motorcycle covers the distance between the two observers $2t$ hours faster than the truck does. The motorcycle is traveling twice as fast as the truck so it covers the distance between the observers in half the time. Therefore the truck must take $4t$ hours to cover the distance between the observers while the motorcycle takes $2t$ hours.

Now the bus passes the first observer t hour ahead of the truck and it passes the second observer $2t$ hours ahead of the truck. The truck takes $4t$ hours to go between the observers so the bus takes $3t$ hours to go the same distance. Consequently, the bus is going $\frac{4}{3}$ the speed of the truck or 40 mph.

Problem 2.9. Suppose we denote the number of digits in 2^{2002} by m, and the number of digits in 5^{2002} by n. We want to find $n + m$. If we multiply the two numbers, we get 10^{2002}, which has 2003 digits. This turns out to be useful because for any positive integer k, 10^k is the smallest number with $k+1$ digits.

Since n is the number of digits in 5^{2002}, we have $10^{n-1} < 5^{2002} < 10^n$. We have used the fact that 10^n has one more digit than 5^{2002}, and so it is larger. Analogously, $10^{m-1} < 2^{2002} < 10^m$. Multiplying the inequalities, we get $10^{m+n-2} < 5^{2002} < 10^{m+n}$, hence $m+n-2 < 2002 < m+n$. Therefore, $m+n = 2003$, and there are 2003 digits on the page.

Solutions to Problem Set 3

Problem 3.1. (a) Yes it's possible. For example, by alternating the baskets with one orange with those with two oranges.

(b) No. If you are having trouble understanding why, see the argument below.

(c) No. Notice that the number of oranges in adjacent baskets must alternate between being odd and even. For example, if the first holds an even number, the second holds an odd number, the third an even number, and so on. If the first basked holds an even number there will be an even number in the ninth basket, the same parity as in the first basket. But these baskets are adjacent! The same contradiction occurs if the number of oranges in the first basket is odd. It is impossible to arrange the oranges this way. The same parity argument applies in (b).

Problem 3.2. To solve this problem, it is not necessary to calculate the sum of all of the even numbers in the first 100, calculate the sum of all of the odd numbers in the first 100, and then compare the two results. Divide the numbers from 1 to 100 into pairs: 1 and 2, 3 and 4, 5 and 6, ..., 99 and 100. There are 50 pairs, and each pair has one odd and one even number, with the odd being one less than the even. So the sum of the odd numbers among the first 100 positive integers is 50 less than the sum of all of the even numbers among these integers.

Problem 3.3. (a) Bob, the second player, can guarantee himself victory, if on each move he moves the marker a different number of spaces than Alice just did on her move. Thus, for every two turns, the marker will move exactly 3 boxes forward. After 8 such pairs of moves, Bob will place the marker into the 25th box, and Alice has no move that stays on the board.

What if there were 26 boxes? Then Alice, the first player, would move the marker one space on her first turn. Then they are playing the same game with 25 boxes, but the second player is now forced to go "first". In this case Alice can guarantee victory for herself.

What if there are 27 boxes, or 50, or 100? Can you answer who will win in each of these cases? If so, here is an extra problem for you. What if the person putting the marker in the last square loses the game?

(b) Bob, the second player, will guarantee himself victory if every time Alice moves the marker 1 box, Bob will answer with a move of 3 boxes, every time

Alice moves 2 boxes Bob will answer with a move of 2 boxes, and every time Alice moves 3 boxes, Bob moves 1 box. Then in every pair of turns the marker moves 4 boxes, and in 6 pairs of turns the second player will reach the 25th box.

How can one discover such a strategy? A player can reason like this. On my turn the 25th box causes a loss for me because, from it, there are no further possible moves. The 24th, 23rd, and 22nd boxes are winners for me because, if the marker is in one of those boxes, I can always move it into the 25th box and win. But if the marker is in the 21st box, then no matter how I move my opponent will be able to move it to the 25th box. So the 21st box is a loss for me. Reasoning similarly, we find the rest of the losing boxes are 17, 13, 9, 5, and 1; they occur with every fourth box. The one who goes first, Alice, moves from the first box and is guaranteed to lose. No matter where she moves the marker, her opponent ends up in a winning box from the start, so the winner is always the second player. Check that this reasoning leads to the strategy for part (a).

Solve the problem in (b) if there are 24 boxes, or 26. What if the one who places the marker in the last box loses?

Problem 3.4. One way to reason would be as follows. Andy can't have more than 1000 books because then the first and third statements would both be true. He can't have between 1 and 999 books because both the second and third statements would be true. That leaves two possibilities, either 1000 books or no books. Check that both are valid solutions.

Problem 3.5. (a) Let us first solve the simpler problem when there are only three coins. How can we find the counterfeit with one weighing? We compare two coins on the scales. If they are balanced, then the coin which was not weighed is the fake. If they do not balance, the lighter one is fake.

Now return to the nine-coin problem. Divide the coins into three piles of three coins each. Compare the weights of the first and second piles. If the scales balance, the counterfeit coin is in the third pile. If they do not balance, the lighter pile contains the fake. With one weighing we found a pile of three coins containing one counterfeit. Now we are back to the simpler problem with three coins and one weighing.

(b) If the solution in part (a) is clear, you should be able to find one fake coin out of 27 with three weighings.

How can we prove that two weighings are not enough? It is not that easy, but here is one way to see it. After the first weighing, we have divided the coins into three groups: the first contains coins that were on the left side of the balance, the second contains coins that were on the right side, and the third contains coins we did not weigh. The number of coins in the groups can be different, and one of the piles might have no coins at all, but, one of the groups will hold at least one-third of the 27 coins or at least 9. The counterfeit might end up in the group with at least 9. With the second weighing, the coins in this group will again be divided into three groups.

Again, whatever our tactics, after two weighings we will have a group of at least three coins that might contain the fake, and no more weighings to sort out which is the fake. So we can't find the fake with only two weighings.

Problem 3.6. (a) Place seven points A, B, C, D, E, F, G in order around a circle; they will represent agents, but for the moment the correspondence between agents and points is unknown. If one agent is watching another, we will draw an arrow from the watcher to his target. We start with no arrows.

We know from the statement of the problem that each agent watches exactly one other, and that each agent is watched by exactly one other. (Why does no agent watch himself?)

We first assign point A to agent 001. The agent watched by 001 can be represented by the next point on the circle, called B. Once this assignment is made, we draw an arrow from A to B. The agent represented by B watches an agent (known to be 002) who is neither himself nor 001, so we let B's target be the next point around the circle, called C, and draw an arrow from B to C. Next, the arrow from C leads either back to A, forming a cycle of three arrows, or to an untouched point, which we can take to be the next point on the circle, D. Likewise the arrow from D either leads back to A or to an untouched point; we cannot have the situation in the figure above, because that would mean that one agent is being watched by two others. So at any step, we either advance one point around the circle, or we go back to A. In any case we get a cycle of agents, each watching the next.

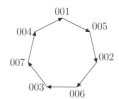

But now we see that this cycle must include *all of the agents*, because, by the statement of the problem, 002 is in the same cycle as 001; 003 is in the same cycle as 002; and so on all the way up to 007. All that remains now is to number every other point around the cycle with increasing agent numbers: A is 001, B is the agent watched by A, so C is 002; D is the agent watched by 002, so E is 003; and so on. We obtain the labeling shown immediately above.

(b) The reasoning in (a) will find any possible arrangement that satisfies the rules, but it breaks down in the case of eight agents (why?). So there is no way of following the instructions with 8 agents.

For what integers n can one follow the instructions with n agents?

Problem 3.7. (a) See the diagram.

(b) No, it's not possible. We will prove it by contradiction. Let's say that there is such a broken line. Count the number of interior points of intersection of the segments of this curve. There is exactly one point of intersection on each segment. There are 7 segments and each point of intersection lies on exactly 2 segments. So how many total points of intersection are there? Exactly half of the total number of the

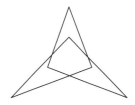

segments or 3.5 points, but this is impossible since there must be a whole number of points. Thus there is no such curve.

Problem 3.8. We can instead consider the problem with 111 matches, taking either 1 or 11 matches per turn. The modified problem will yield the same strategy. After each of Jack's turns, the number of matches will be even, and after each of Jill's turns it will be odd. Therefore, no matter how long they play, Jill can't win because after any of her turns the resulting number of matches cannot be 0. And Jack will always win since Jill will always leave at least one match and Jack can still play.

Problem 3.9. (a) Imagine that the hallway extends from left to right. Let some part of the floor be covered by three (or more) rugs. From the rugs that cover that part, keep the one whose left end is the farthest left and the one whose right end is the farthest right. If these are the same rug, keep that one rug. Then we can take away the other rugs that cover this portion of the floor because the portion of the floor we are considering will be covered by the two rugs we kept. By taking away extra rugs, we will reach a state where any part of floor is covered by at most two rugs.

(b) From part (a) we can assume that no more than two rugs cover any portion of the floor. We can also remove any rug that is completely covered by or totally on some other rug. Now label the rugs consecutively, starting from the one that is farthest left. Now there is only one rug that reaches the left side of the hallway, and it will be 1. There is exactly one rug that touches it or covers some part of it, and it will be 2, and so on. With this method of numbering, rugs with odd numbers do not intersect each other, and rugs with even numbers do not intersect each other. But all the rugs together cover the whole floor. If the even-numbered rugs cover less than half of the hallway, the odd-numbered rugs will cover more than half the hallway. Thus we can remove either all odd-numbered or all even-numbered rugs so that the remaining rugs cover at least half the hallway.

Problem 3.10. If we subtract the same number of pounds from the weight of each cow, the assumptions of the problem do not change because we can still remove each cow and divide the remaining ones into two groups of 50, each of which weighs the same. Now choose the lightest cow and subtract its weight from the weight of each of the remaining cows. We will end up with one cow that weighs 0 pounds and another 100 cows. The total weight of the remaining 100 cows is an even number because we can divide them into two groups of equal weight. Is there a cow whose weight is an odd integer? If there were, we could take it out and put in the one weighing 0. Then the total weight of the 100 cows would be odd, and they could not be sorted into two groups of equal weight. Therefore, each of the cows weighs an even number of pounds. If any of the cows had a nonzero weight we could change the problem again by dividing all the nonzero weights by 2. We can apply the argument above to again conclude that there cannot be a cow whose weight is odd. Divide by 2 again, and so on. We can't divide the weights by 2 forever. Thus, all of the weights must have been 0, and all of the cows weigh the same.

Problem 3.11. The last digit of E^3 is E itself, because E×E has last digit F and E×F has last digit E. Thus E must be 4 or 9 (the cases 0, 1, 5, 6 are excluded because they lead to F = E).

Next, 9 is excluded because it would lead to F = 1, and the last line above the second stroke would only have 4 digits instead of 5. So E = 4 and F = 6.

Since E is even, both V and I have to be even (why?) and they cannot be 0 (why?). Therefore either I = 2 and V = 8, or vice versa. It is easy to see both cases work out, so the number is either 6284 or 6824.

Solutions to Problem Set 4

Problem 4.1. (a) It is possible to get from A to B in 5 different ways depending on which road we choose. For each of those, there are 4 ways to get from B to C. Thus we get the total of $5 \times 4 = 20$ ways to get from A to C by going through B without visiting any other city.

(b) To get from A to C via D, you can go $2 \times 3 = 6$ different ways. Thus, to get from A to C through B or D, you can go $20 + 6 = 26$ different ways.

Problem 4.2. Let us solve (b), which will also solve (a). To measure exactly 15 minutes, start the 7-minute and 11-minute hourglasses simultaneously. When the 7-minute hourglass runs out, flip it over. After another 4 minutes, when the 11-minute hourglass runs out, flip over the 7-minute hourglass again. The sand in this hourglass will drop for another 4 minutes. Thus, we have measured $7 + 4 + 4 = 15$ minutes.

Problem 4.3. Suppose there are x people in the country who are both mathematicians and musicians. How many mathematicians are there in total? There must be $20x$ mathematicians because this is the only way that every 20th mathematician can also be a musician. What about musicians? In the same way we find there are $30x$ of them. Thus there are 1.5 times more musicians than mathematicians in the country.

Problem 4.4. (a) When only one bulb is lit, there are three options. When exactly two are lit, exactly one is not lit, and there are three ways this can happen. Finally, there is the event that all three lights are lit. Thus there are seven possible ways to have some of the lights on.

(b) Solution 1: We can try to count the options as we did in (a). When only one bulb is lit, there are five options; when four are lit, there is exactly one not lit, so there are also five options.

How can we count the options when exactly two bulbs are lit? One way is like this. Lay out the bulbs in a row. If the two lit bulbs are adjacent, there are four possibilities: $**---$, $-**--$, $--**-$, and $---**$. If they are separated by one unlit bulb, there are three options: $*-*--$, $-*-*-$, $--*-*$. If they are separated by two unlit bulbs, there are two possibilities: $*--*-$ and $-*--*$; If the lit bulbs are three apart, the only option is $*---*$. Thus there are $4+3+2+1 = 10$ options for two bulbs. There are an equal number of options for three bulbs because, when three bulbs are lit, two are off. There is one more option in which all the bulbs are on. Thus the number of possibilities is $5 + 5 + 10 + 10 + 1 = 31$.

Solution 2: The previous solution turned out to be complicated. What if there were 10 bulbs? We would have to look at a large number of options. Is it possible to make the solution simpler? Let us count the ways the bulbs could be on and off including the possibility that no bulb is lit. With three bulbs there are 7 possibilities with at least one bulb lit and one with no bulb lit for a total of 8 ways. Add one more bulb to the string. When the additional bulb is on, there are 8 options for lighting the other three bulbs, and when the additional light is off, there are also 8 options for the other bulbs. This gives a total of $8 + 8 = 16$ options for a string of four lights. Of these 16, one option is all bulbs off so there are $16 - 1 = 15$ options with at least one bulb on.

Using this approach, we can progressively solve the problem for any number of bulbs. See if you can use this approach to find the solution for 10 bulbs.

Problem 4.5. To see the pattern, take $n = 5$ as an example and draw $5^2 = 25$ points in a 5×5 square. We can also count the points in a different way by counting along the diagonals that run down and to the right at 45 degrees. Start at the point in the lower left of the square; there is 1 point. On the next diagonal up and to the right there are 2 points. On the third diagonal there are 3 points Continue to the fifth diagonal with 5 points. The sixth diagonal has 4 points; the seventh has 3, and so on to the last with 1. Therefore we have $1 + 2 + 3 + 4 + 5 + 4 + 3 + 2 + 1 = 5^2$. If we draw a square with n^2 points and count along diagonals as above, we will see that $1 + 2 + \cdots + (n-1) + n + (n-1) + \cdots + 1 = n^2$.

Problem 4.6. (a) First, let's agree that we will count the word itself as one of its anagrams.

Solution 1: The given word has 7 letters, and these letters are distinct. To count the number of anagrams of a word with exactly one letter is simple; there is exactly one anagram. To the one-letter word, let's add a second letter different from the first. It can be placed either in front or at the back of the given letter, so there are two anagrams of a two letter word. Add one more different letter. We have to take all existing strings of letters and add the new letter to them. For every given string of two letters, it can be done in three different ways by putting the new letter in the first, second or third place. Therefore a 3-letter word will have $2 \cdot 3 = 6$ anagrams; the fourth letter can be added at 4 different places, so a 4-letter word has $4 \cdot 6 = 24$ anagram; a 5-letter word, $5 \cdot 24 = 120$; and so on. For a 7-letter word we get $2 \cdot 3 \cdot 4 \cdot 5 \cdot 6 \cdot 7 = 5040$ anagrams. The product of all consecutive integers from 1 to n and is denoted by $n!$, which is read as "n factorial". Thus the number of anagrams of a word with 7 distinct letters can be written as $7!$.

Solution 2: Let us write all the anagrams on 7 sheets of paper. The first sheet will contain all anagrams with the first letter R; the second has all with the first letter E; and so on, so that the last sheet will contain all anagrams whose first letter is Y. Each sheet will have exactly the same number of anagrams since one letter can always be exchanged for another. Thus in order to find the total number of anagrams, we can count their number on one sheet, and then multiply it by 7. Let's now take all the anagrams on the first sheet. That number is equal to the total number of anagrams of the word EALSPY. By the same argument, this number is 6 times the number of anagrams of the word ALSPY. Thus, to find the total number of anagrams, we can count the number of anagrams of the word ALSPY, and multiply it by 7 and by 6. Continuing in the same fashion, we'll get down to the word consisting of just the letter Y. This word has only one anagram, so we must multiply 1 by $7 \cdot 6 \cdot 5 \cdot 4 \cdot 3 \cdot 2$. We get the final answer of $7! = 5040$ anagrams.

(b) Parsley; players; replays, sparely, parleys.

(c) SOLVE MATHEMATICAL PROBLEMS DAILY.

(d) APPLE is a five-letter word, but it is made up of only four different letters as P appears twice. Let's make the two P's temporarily different by leaving one of them a capital and making the other one lowercase, so the word becomes APpLE. How many rearrangements does this word have? The number is $5! = 120$ following the reasoning in part (a). These can be paired off by interchanging P with p, so APpLE and ApPLE form a pair as do PLEAp and pLEAP. When we return the lowercase letter to a capital, both "words" in a pair give the same rearrangement of APPLE. There are $120/2 = 60$ of these pairs, so there are 60 "words" that can be formed by rearranging the letters of APPLE.

(e) For BAOBAB, we can temporarily make the three B's and the two A's different as in BAObaB. There are 6! ways of rearranging BAObaB. When we make the B's the same, all arrangements of the three different B's give the same rearrangement of BAOBaB, and there are 3! rearrangements of the three different B's. Thus there are $6!/3!$ different rearrangements of BAOBaB. Finally make the A's the same to get

$$\frac{6!/3!}{2!} = \frac{6!}{3! \times 2!} = 60$$

ways of rearranging BAOBAB.

Problem 4.7. (a) If the lines intersect, the plane is divided into 4 parts. If the lines are parallel, the plane is divided into 3 parts:

(b) If the three lines are all parallel, the plane is divided into 4 parts. If two of the lines are parallel and the third intersects them, the plane is divided into 6 parts. If no two lines are parallel, look at two intersecting lines. The third line can either go through their point of intersection or not. If there is one common point of intersection, the plane is divided into 6 parts. If not, the plane is divided into 7 parts. Thus three lines can divide the plane into 4, 6 or 7 parts.

(c) As in (b), separate into cases by the position of the lines. If all lines are parallel, they separate the plane into 5 parts. If three are parallel and the fourth is not, there are 8 parts. If two of the lines are parallel and neither of the remaining two are parallel to the first two, there still are several cases. If the two other lines are parallel to each other, there are 9 parts. If the other two lines intersect, then either their intersection point lies on one of the parallel lines and there are 9 parts, or it does not and there are 10 parts:

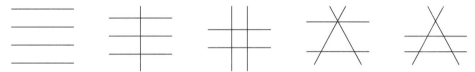

Finally, if there are no parallel lines, we have several possibilities: all lines intersecting in one point, giving 8 parts; three lines intersecting in one point and the fourth line not going through that point, giving 10 parts; and no three lines intersecting in one point, giving 11 parts. Thus four lines can divide the plane into 5, 8, 9, 10 or 11 parts.

A general question is this: For every natural number n, into how many parts can n lines divide the plane? There is no known exact answer to this question.

Problem 4.8. Draw the lines representing the covered edges of the sheet. Also draw the line MN which cuts the rectangle $ABCD$ into two rectangles. Rectangle $MBCN$ is exactly half covered while rectangle $AMND$ is more than half covered, as one can see by drawing the diagonal MD. Therefore the uncovered part of the sheet is smaller than the covered part.

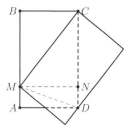

Problem 4.9. The diagram below shows we can put 4 beans in each column so the number of beans in successive rows is 3, 2, 1, 0, 5, 6, 7 and 8, all different numbers.

How could one come up with this solution? Since each row can contain from 0 to 8 beans, there are 9 possibilities in all. Since each row is to contain a different number of beans, one must choose eight of these possibilities and reject one. How can one decide which one to reject? The total number of beans must be divisible by 8 since there are eight columns, and each column has the same number of beans. Now $0+1+\cdots+8 = 36$, and we must leave out one of these numbers and get a sum that is a multiple of 8. The only number that does this is 4. Now the rows must contain 0, 1, 2, 3, 5, 6, 7, and 8 beans, and, because there are 32 beans total, each column must contain 4 beans. Given these considerations, see if you can find a suitable arrangement.

Problem 4.10. (a) Place a pyramid in front of you with the yellow face down and turn it so the green face is toward you. Now there are two "rear" faces which, from left to right, can go red-blue or blue-red. Therefore there are two different pyramids.

(b) Imagine one of each of the different kinds of cubes are on the table in front of you. Place all of the cubes with the yellow face down and divide the cubes into five piles with each pile having the same color face on top. There will be an equal number of cubes in each pile so we can count the number of cubes in one pile and multiply by 5 to find the total number of cubes.

Turn the cubes in the pile with red tops so the green faces are toward you. The rear face can be blue, white or black. For each of these three possibilities, the remaining two colors can either be on your left or right. For example if the rear face is blue, the left face can be either black or white. This gives a total of $3 \times 2 = 6$ cubes in the pile with red face up. Therefore there are $5 \times 6 = 30$ cubes on the table.

Solutions to Problem Set 5

Problem 5.1. See diagram on the right. A quadrilateral with four equal sides is called a rhombus. It is a square only when all of its angles are also equal.

Problem 5.2. (a) We are assuming here that one student cannot be a hall monitor and the president at the same time. There are 25 ways to choose a hall monitor. For each of these choices, there are 24 ways to choose the president. Hence we get $25 \times 24 = 600$ different ways.

(b) Let's use the previous part. In each pair found in part (a), let the president become a hall monitor. Then we get all possible pairs of hall monitors, but each pair is counted twice because if Will and Simon are the hall monitors, we will obtain this pair from both the choice "Will is hall monitor, Simon is president", and from the choice "Will is president, Simon is hall monitor". Therefore the number of choices is half as many, or 300.

(c) The first hall monitor can be chosen 25 different ways. After the first has been chosen, the second can be chosen in 24 different ways. After the first two have been chosen, the third can be chosen in 23 different ways. If we keep track of the order in which the monitors were chosen, there are $25 \times 24 \times 23 = 13\,800$ ways of doing this. But we do not care what order they are picked in, only that there are three monitors. If Freddie, Tran and Jaz are chosen as monitors, this could have happened with Freddie first, Tran second, Jaz third; with Freddie first, Jaz second, and Tran third; with Tran first, Jaz second, Freddie third; with Tran first, Freddie second, Jaz third; with Jaz first, Freddie second, Tran third; or with Jaz first, Tran second, Freddie third. We have counted each set of hall monitors six times in our process, so the total number of ways of choosing three hall monitors is $13\,800/6 = 2300$.

Problem 5.3. Fold the ribbon in half lengthwise. The resulting twofold ribbon is 72 inches long. Folding it again results in a 36-inch fourfold ribbon. Fold it twice more to get a thick 9-inch ribbon. Now unspool three 9-inch lengths from the ribbon. The result is a 27-inch piece of ribbon.

Problem 5.4. Suppose that all of the children have gathered different numbers of nuts. Write, in ascending order, the amounts each has gathered. The first will be at least 0, the next at least 1, and so on. The last is at least

14. Therefore the children gathered at least $0 + 1 + 2 + 3 + \cdots + 13 + 14$ nuts. One way to calculate this sum is $1 + 2 + \cdots + 13 + 14 = (1 + 14) + (2 + 13) + (3 + 12) + \cdots + (6 + 9) + (7 + 8) = 7 \times 15 = 105$. But 105 is greater than 100, so our assumption was incorrect and two or more children gathered the same number of nuts.

Problem 5.5. (a) There are $47 - 23 = 24$ employees who speak Spanish but not German, and $35 - 23 = 12$ who speak German but not Spanish. Thus $24 + 12 + 23 = 59$ know at least one of the two languages, while $67 - 59 = 8$ know neither language.

(b) First let's find the number of employees who speak at least one of the three languages. We can start by adding the numbers of employees who speak Spanish, German, and French. The result is $47 + 35 + 20 = 102$. We have counted everyone who speaks one of these languages, but employees who speak two of these languages have been counted twice, and those who speak all three have been counted three times. Let's adjust the total by subtracting the number of those who speak Spanish and French, who speak French and German, and who speak Spanish and German. The result is $102 - 23 - 12 - 11 = 56$. Each employee who speaks all three languages has been subtracted three times after they were initially added in three times, so they are not now in the count. Therefore, we now need to add them in, so the number of employees who speak at least one of the three languages is $56 + 5 = 61$. Therefore the number of employees do not speak any of the three languages is $67 - 61 = 6$.

Problem 5.6. (a) Yes. If the first switch turns on the first light, the second turns on the second and third lights simultaneously, and the third switch controls the remaining four lights, you can check that, by changing the positions of the switches, we can light up any number of bulbs from 0 to 7.

(b) How many total distinct switch positions are there? Each of the three switches can be either on or off, so the total number of possible positions is $2 \times 2 \times 2 = 8$. This means that the lights can't be lit in more than 8 different ways that correspond to the 8 possible switch positions. But the number of bulbs that we would like to be able to turn on are the numbers from 0 through 8. This means there need to be 9 total positions, which is impossible.

Problem 5.7. Yes, because the lengths of the runners are identical. To find the length of a runner, add the sum of the lengths of the vertical parts of one staircase to the sum of the horizontal parts of the same staircase. For both of the staircases, the sum of the horizontal segments must equal the total base length of 2 meters, while the sum of the vertical segments must equal the height of 1 meter. Therefore the length of each runner is the 3 meters, and each runner will cover either staircase.

Set 5

Problem 5.8. The diagram shows that a 60-degree range on the left thermometer has the same length as a 100-degree range on the right thermometer. Thus a three degree range on the left is the same length as a five degree range on the right. Let the heights of the mercury columns be equal at x degrees. Then the range of $x - 10$ degrees on the left thermometer has the same length as an x degree range on the right. It follows that
$$\frac{x-10}{3} = \frac{x}{5},$$
and so $x = 25$. The mercury levels are at the same height at 25 degrees.

Problem 5.9. (a) Label the books 1, 2, and 3. At first they are in their original positions of 1, 2, 3, on the shelf. We can place book 1 in position 2, and then book 3 must go to 1, so the only possible result is 3, 1, 2. If we place book 1 in position 3, then book 2 must go in position 1, so the only possible result is 2, 3, 1. Thus there are two ways to arrange the books with none in its original position.

(b) Label the books 1, 2, 3, and 4. At first they are in the original positions of 1, 2, 3, 4 on the shelf. We can place book 1 in position 2, 3, or 4. With each of these moves, we will find the same number of options for arranging the books. Therefore let's count the number of arrangements when book 1 ends up in position 4, and then multiply that number by 3 to get the total number of arrangements. Place book 4 temporarily in position 1, so we now have the books in the order 4, 2, 3, 1. The two rearrangements of the first three books, now 4, 2, 3, that leave none in its current place, will give an arrangement of all four books that leave none in its original place. In addition, the arrangement 4, 3, 2 gives an arrangement of 4, 2, 3 that leaves none of the four books in its original place. There are three arrangements of 4, 2, 3 that lead to arrangements of all four with none in its original place with book 1 in the fourth place. Multiply this number by 3 to get a total of nine rearrangements that leave no book in its original position.

(c) Label the books in order 1, 2, 3, 4, 5. Book 1 can be placed in any one of four places: 2, 3, 4, or 5, and the number of possible rearrangements in each of these cases will be the same, so we can calculate the number of outcomes with book 1 in position 5 and multiply by 4. Place the book 5 temporarily in position 1. Then books 5, 2, 3, 4 should be rearranged either so that book 5 remains in position 1 while the books 2, 3, and 4 are rearranged amongst themselves, or so that no one of these four books remains in its position. From (a) we know there are two arrangements with book 5 occupying the first position, and from (b) we know there are 9 arrangements when we move book 5. This gives 11 rearrangements with book 1 in position five, so there are $4 \times 11 = 44$ arrangements of five books with none occupying its original position.

Problem 5.10. First we will consider the greatest odd divisor for each of the numbers from 1 to 50. For example, the greatest odd divisor of 16 is 1, the greatest odd divisor of 20 is 5, and that of 49 is 49. The possible odd divisors of the first 50 positive integers are $1, 3, \ldots, 49$, and there are 25 of them. Because we have chosen 26 numbers from 1 to 50, there must be at least two that have the same greatest odd divisor. Then one of those two numbers will be divisible by the other. You may need to think about why the previous sentence is true.

Solutions to Problem Set 6

Problem 6.1. A rhombus has four equal sides. Draw a diagonal equal in length to a side and the rhombus has been divided into two equilateral triangles. Each triangle has three 60-degree angles, so the rhombus has two 60-degree angles and two 120-degree angles.

Problem 6.2. We can write $\frac{2005}{2006} = 1 - \frac{1}{2006}$ and $\frac{2006}{2007} = 1 - \frac{1}{2007}$. This means that the second fraction is greater than the first.

Problem 6.3. To find the last digit in the square of an integer, it is enough to know just the last digit of the number itself. We build a table:

If a number ends in 0, its square ends in 0.
If a number ends in 1, its square ends in 1.
If a number ends in 2, its square ends in 4.
If a number ends in 3, its square ends in 9.
If a number ends in 4, its square ends in 6.
If a number ends in 5, its square ends in 5.
If a number ends in 6, its square ends in 6.
If a number ends in 7, its square ends in 9.
If a number ends in 8, its square ends in 4.
If a number ends in 9, its square ends in 1.

We have examined all possible cases and see that there is no case where the square ends in 2.

Problem 6.4. We have already solved similar problems. Can you think of which ones are similar?

Mark 25 points on a circle to represent students in the class. Connect pairs of friends with line segments. If every student is friends with 7 classmates, 7 line segments would leave each point, so there would be $\frac{25 \times 7}{2}$ line segments. The 2 in the denominator is there to avoid counting each friendship twice. This would give us a fractional number of segments, which is impossible.

Problem 6.5. We can see that the position of words that occur in more than one verse is not the same in the two languages — apparently YumYum and English have different rules about the order of words in a sentence. Let's try to figure out which words correspond to which.

There is only one word appearing in all three English lines: cat. In the YumYum version there is also a unique word with this property, ту. Therefore ту in YumYum means cat.

The last two lines of the poem share one other word besides cat/ту: in English it is mouse, and in YumYum it is ля. Hence ля means mouse.

In the first two lines, too, there is one recurring word besides cat: it's giant in English and ам in YumYum. This means that ам = giant. Applying the same argument to the first and third lines, we see that ям = scary. Finally, by elimination, we find that му = was, бу = saw, and ру = ate.

Problem 6.6. Any natural number ends in one of 10 digits. Because there are 11 numbers, two of them will have the same one's digit. Therefore their difference will end in zero and so will be divisible by 10.

Problem 6.7. Let's say Carol drank orange juice for x minutes. That means she looked out the window $2x$ minutes, slept $4x$ minutes, and read her book $8x$ minutes. Carol spent a total of 1 hour on everything, so

$$x + 2x + 4x + 8x = 60.$$

This gives $15x = 60$, so $x = 4$. Therefore Carol began looking out of the window $8x + 4x = 48$ minutes after noon, or at 12:48 p.m.

Problem 6.8. Suppose we could not dig such a tunnel. That would mean that the point diametrically opposite every point on land is sea. Therefore, there would be at least as much sea as land. But land covers more of Urth than sea, so there is more land than sea. This is a contradiction, so the assumption was false and we can dig such a tunnel.

Problem 6.9. In fact, you don't even need to use any blank rectangles; see the diagram.

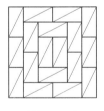

Problem 6.10. First let's try to solve the problem when the number of digits in the string is small. We will call the number of digits in the string the length of the string. For example, 0 and 1 are the only strings of length 1. There are two of them and both satisfy the problem's requirements. The strings of length 2 that satisfy the requirements are 01, 10, and 11, and there are three of them.

Let us analyze how a string of length n with no adjacent 0's is built. If there is a 1 in the last place of the string, then in the previous $n-1$ positions there can be any string without adjacent 0's. If there is a 0 in the last place of the string, then there must be a 1 in the previous place and any string without adjacent 0's in the $n-2$ positions before that. Therefore any string of length n with no adjacent 0's results from either the addition of a 1 to a string of length $n-1$ with no adjacent 0's, or by the addition of 10 to a string of length $n-2$ with no adjacent 0's.

Now we can systematically find the number of strings with adjacent 0's of greater and greater length by knowing the number of such strings of smaller length. For example, for $n = 3$ there are $2 + 3 = 5$ strings of length three. This is so because we can add 10 at the end of the two strings of length 1, giving 010 and 110, and can add 1 at the end of the three strings of length 2, giving 011, 101 and 111. Similarly, there are $3 + 5 = 8$ such strings of length 4, and so on. The resulting sequence, where each number beyond the second is the sum of the two preceding numbers, is known as a Fibonacci sequence. For this problem the sequence runs $2, 3, 5, 8, 13, \ldots$, and you can extend it to find the answer for strings of length 15.

Problem 6.11. The answer is no. The longest segment that can be covered by an equilateral triangle is no longer than the side of that triangle. We have one big equilateral triangle and two smaller ones. One of the small triangles cannot cover two vertices of the larger one. Therefore two small triangles can cover at most two vertices of the big one, so one vertex will remain uncovered.

Solutions to Problem Set 7

Problem 7.1. The train is moving at a rate of 2 meters per second. The front of the train takes 360/2=180 seconds to cross the bridge. However the train must also leave the bridge, so its end must cover another 180 meters which takes another 90 seconds. Therefore the train will pass the bridge in 270 seconds.

Problem 7.2. See diagram. The dashed lines represent the boundary of the original pond, while the solid lines show the boundary of the new one.

Problem 7.3. (a) If each student had a birthday in a different month, we could have at most 12 students. There are 25 students, so at least two have their birthdays in the same month.

(b) The answer is yes. If each month had at most two birthdays, there could be at most 24 students, but 25 are in the class.

Problem 7.4. (a) Make a table of remainders when powers of 3 are divided by 5. The remainder for 3^1 is 3. Since $3^2 = 9$, the remainder is 4. Since $3^3 = 27$ the remainder is 2. The remainder for $3^4 = 81$ is 1. After this, the remainders will repeat in the same pattern, as we see with $3^5 = 243$, which has a remainder of 3 and $3^6 = 729$ for a remainder of 4. The remainders occur in the pattern $3, 4, 2, 1, 3, 4, 2, 1, \ldots$. Since 100 is a multiple of 4, the pattern cycle will be at 1, so the remainder of 3^{100} on division by 5 is 1.

To calculate remainders, it is convenient to use a general theorem which is easiest to introduce with examples. The remainder on dividing 78 by 5 is 3, and the remainder on dividing 79×5 is 4. The remainder on dividing $157 = 78 + 79$ by 5 is the remainder on dividing $3 + 4 = 7$ by 5, which is 2. The remainder on dividing $6162 = 78 \times 79$ by 5 is the remainder on dividing $3 \times 4 = 12$ by 5, which is 2. It does not matter what number you are dividing by to get the remainders, it does not matter what numbers are being divided, and it works with addition, multiplication, and, when correctly understood, subtraction. When subtracting, one can get negative numbers, but the remainder is still supposed to be nonnegative. The remainder when -8 is divided by 5 is 2 because $-8 = (-2) \times 5 + 2$.

(b) Put together a table of remainders when powers of 5 are divided by 3. Use the theorem above, replacing the numbers by their remainders when divided by 3. The remainder for 5^1 is 2, and for $5^2 = 25$ it is 1. The remainder when dividing $5^{100} = (5^2)^{50}$ by 3 is the remainder on dividing 1^{50} by 3, which is 1.

Problem 7.5. The answer is yes. Split the numbers into pairs: $(1,2)$, $(3,4)$, $(5,6)$, ..., $(49,50)$. The difference between the numbers in each pair is 1. There are 25 pairs. Because we have chosen 26 numbers, we will definitely have at least two numbers from one pair and they will differ by 1.

Problem 7.6. (a) The top left triangle consists of two halves of a square, hence its area is 1. The bottom left triangle lies inside a 2×2 square outlined by dashed lines, as seen on the right. To find its area, we need to subtract the areas of the extra pieces of the square. One of these pieces is half of the 2×2 square and the second extra piece is a half of a 1×2 rectangle. Thus the area of the triangle is $4 - 2 - 1 = 1$. The triangle on the right lies inside a 2×3 rectangle outlined by dashed lines on the diagram. Subtracting the areas of extra pieces, we conclude that the area of the triangle is 2.

(b) Our triangle lies inside a 4×3 rectangle outlined by the dashed lines in the diagram on the right. By subtracting the areas of extra pieces, we get the area of the triangle as

$$\frac{3 \times 4}{2} - 1 - \frac{3 \times 1}{2} - \frac{2 \times 1}{2} = \frac{5}{2}.$$

(c) Solve this problem on your own using procedures similar to the ones described in parts (a) and (b). This can be done in many different ways, but the answer is 13.

Problem 7.7. (a) The number's first digit can be either 1 or 2 which gives two possibilities. For each choice of first digit, there are two choices for second digit, and so on. Therefore there are $2^{10} = 1024$ such numbers. This argument may remind you of other problems you have worked on.

(b) For this problem we don't care about the order in which the mangoes are picked, but only which mangoes have been picked. The total number of mangoes picked can be anywhere from 1 to 10. We need to count all of the possibilities.

We can make this problem similar to the previous one. We can either pick or not pick each fruit. Let us write down each way of picking mangoes as a string of ten digits with each digit matched to a particular mango. A digit will be a 1 if that mango is picked and 2 if it is left on the tree. Therefore each way of picking mangoes will correspond to a ten-digit number made up of 1's and 2's, and every such ten-digit number will correspond to a way of picking mangoes. The one exceptional case is 2222222222 which corresponds to picking no mangoes at all. Therefore, there is one less way to pick several fruit than there are ten-digit numbers made up of 1's and 2's. This gives us $1024 - 1 = 1023$ ways of picking anywhere from one to ten mangoes. We could also consider the possibility of not picking any fruit by counting it as a way for picking zero mangoes. Then the answer would be 1024 ways of picking mangoes.

Problem 7.8. Draw the rectangle on the graph paper with the long side horizontal. Draw a diagonal from vertex A at the lower left to vertex C at the upper right. The rectangle consists of three horizontal 200×1 strips, each with vertical grid lines. The diagonal first follows the lower strip, then the middle, then the top one. We can argue by contradiction that, except for the points A and C, the diagonal does not pass through any lattice point, a point where a horizontal grid line crosses a vertical grid line.

Suppose, contrary to what we want to show, that when crossing from the lower strip to the middle one, the diagonal crosses through  the lattice point X. Then the diagonal connects vertex A with lattice point X. As the diagonal continues, it will cross from the middle strip to the top strip at a lattice point Y. It does this because, in going from A to X, it goes a whole number n of steps horizontally while going one vertically, and, over the next strip, it will also go n steps horizontally in going 1 vertically. Finally in going from Y to C, the diagonal will go n steps horizontally. In going from A to C the diagonal would then go $3n$ units horizontally for some integer n. But this would say the length of our rectangle is divisible by 3, which is false.

Therefore our diagonal does not pass through a lattice point when crossing from the lower strip to the central one. Similarly, it does not pass through a lattice point when crossing from the central strip to the upper one, so there are no lattice points on our diagonal except for A and C. Now, leaving out these endpoints, the diagonal crosses the two horizontal borders of the middle strip and 199 vertical gridlines. Because there are no lattice points between A and C, we end up with a total of 202 intersections with the gridlines other than A and C. These points, along with A and C, separate the diagonal into 203 line segments. Each of these segments is a piece of one box through which the diagonal passes. Therefore the diagonal passes through 203 boxes.

Problem 7.9. Usually friendship is mutual, so if Ashok is Natasha's friend, then Natasha is Ashok's friend, and we just say that they are friends.

Suppose there are n students in the class. Each of them can have from 0 to $n-1$ friends, a total of n options. But there are n students, so if all of them have a different number of friends, there must be a student with 0 friends, a student with 1 friend, and so on to the student with $n-1$ friends. Now there can't be a student with 0 friends in this class along with a student with $n-1$ friends, because the student with $n-1$ friends is friends with all of his classmates, including the one with 0 friends. This is a contradiction, so at least two of the students have an equal number of friends in the class.

Problem 7.10. We solve (a) and (b) together. Suppose the board has its long edge horizontal. With its first move the rook retreats left or right to the edge of the board that is as far as possible from the bishop. Next the rook moves to the central horizontal row, if not already there. In the next move, it either captures the bishop if the bishop is in the central row, or it moves along the central row until, as in the diagram, there is exactly one unoccupied column between the rook's column and the bishop's column.

The bishop will be captured if it moves either to the free column or into the rook's column, so it must retreat away from the rook, while avoiding the central strip. Then the rook moves in the central strip two squares toward the bishop, so the bishop ends up in the same situation as before, and has to retreat again in the same direction. But the board is finite, so the bishop will eventually be unable to move away to avoid capture.

Problem 7.11. Notice that the grasshopper will never be able to reach inside a central circle of diameter 1 meter, because the distance from any point on the patio to any point within that central circle is less than 2 meters. Now we will prove that he can reach any other point of the patio.

First we show that, starting at a point A on the edge of the patio, the grasshopper can reach any other point on the edge of the patio.[1] Suppose that point B is on the edge and close to A. Look at the diagram to see that the circles of radius 2 about A and B will intersect at some point C on the opposite side of the central circle but inside the patio. The grasshopper can then jump from A to this point of intersection to B. Now, in a series of hops, he can get from A to any other point on the edge of the patio.

Now we show that the grasshopper can get to any point D in the ring of width 1 m between the edge of the patio and the central circle of diameter 1 m. Draw the circle of radius 2 m centered at D, as shown on the right. It intersects the edge of the patio either at two points or in one point (if D lies on the inner boundary of the ring). Let B be one of these intersection points. Then the grasshopper first gets to point B on the edge and then hops to D.

---

[1] Using the Pythagorean Theorem, you can even show he can do this in two jumps.

Problem 7.12. This can be a difficult problem, but there is a short and elegant solution. Suppose that the problem's conclusion is false; that is, that no two representatives of one of the four races sit together. Then remove all elves and dwarves. There will be exactly one empty chair between any two remaining representatives, which means that there must be an even number of chairs around the table. But this is false, since there are 2003 chairs.

Solutions to Problem Set 8

Problem 8.1. See diagram on the right.

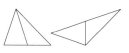

Problem 8.2. It might seem that the answer is 2.5 hours, but it is not. If we start with two bacteria, the dish will get covered in 4 hours and 50 minutes.

How can we see this? The dish that starts with one bacterium fills up in 5 hours. Ten minutes into this growth, there are two bacteria on the dish—the same situation as the dish that starts with two cells. What's left of the 5 hours is 4 hours and 50 minutes, so that's the amount of time needed for a dish to fill up if it starts with two bacteria.

Problem 8.3. Let's divide the square into 16 squares that are 1 cm on a side as in the diagram. These squares do not overlap, so a single dot cannot lie inside two different squares. Since there are 15 dots and only 16 squares, there is at least one square with no dot inside.

Problem 8.4. The problem has a simple solution: add one more goose to the flock—say a gray goose, so it doesn't get confused with the others. With the extra goose, exactly half of the geese will land at each lake. At the seventh lake exactly half of the geese that have got this far will land, and only our gray goose will continue flying; so 1 goose lands. At the sixth lake 2 geese landed; at the fifth lake, 4 geese; at the fourth lake, 8 geese; at the third, 16 geese; at the second, 32 geese; and at the first, 64 geese. Therefore, the total number of white geese is $1 + 2 + 4 + \cdots + 64 = 127$.

Can you think of a simple way to calculate the sum $1 + 2 + 4 + \cdots + 64$?

Problem 8.5. (a) We can take the four points to be vertices of a square. Check that this works.

(b) One solution is the vertices of a regular pentagon, as in the diagram on the right. All sides and all angles of this pentagon are equal. Let's prove that all its diagonals are also equal.

Each diagonal cuts off a triangle with two equal sides—the sides of the pentagon. The included angle is the angle of a regular pentagon, so the included angles for any two of these triangles are equal. The diagonals cut off congruent triangles, and therefore the diagonals are congruent to one another since they are bases of the congruent triangles. This construction is shown on the right.

Now it is easy to prove that any three vertices form an isosceles triangle. Each of the three sides of the triangle is either a side or a diagonal of the pentagon, and so two of them are equal to each other.

(c) The six vertices of a regular hexagon would not work. The triangle formed from two adjacent vertices and a third vertex across the center from one of the others is not isosceles. Yet one can get a suitable arrangement using the picture from part (b) by adding the center point of a regular pentagon. Since the distances from this point to all the vertices of the pentagon are all the same, the new triangles that have the center as one of their vertices are also isosceles.

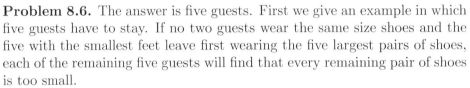

Problem 8.6. The answer is five guests. First we give an example in which five guests have to stay. If no two guests wear the same size shoes and the five with the smallest feet leave first wearing the five largest pairs of shoes, each of the remaining five guests will find that every remaining pair of shoes is too small.

How can we prove that it is impossible to have six guests who must stay? Let's assume that six guests didn't find shoes that fit. Then at the end of the process there are six guests in the house and six pairs of shoes, owned of course by six people. Since $6+6 > 10$, the set of guests that stay and the set of owners of shoes that stay overlap. That is, at least one guest is left in the house together with his or her own shoes! Since those shoes were there when that guest tried to leave, we reach a contradiction: the guest would have been able to leave after all.

Clearly, the same contradiction is reached if we assume that more than six guests stay the night. So the maximum number of guests who stay is 5.

Problem 8.7. A solution can be seen in the diagram. The original position of the top mirror is depicted by a thin line. Reflect the dashed segments representing part of the original path of the light, across this thin line. We will obtain parts of the new path. The length

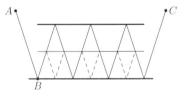

of the new path is the same as that of the original path, and the ray will still get to point C.

Problem 8.8. This problem has a surprisingly short and elegant solution. Let's concentrate on the central cube. It must be cut out, and it has 6 faces, so there must be a cut along each of these faces. Therefore there must be at least 6 cuts! You can cut the big cube into 27 little cubes without rearranging the pieces by making six cuts each of which is parallel to and 1 inch away from each of the six faces of the big cube.

Problem 8.9. (a) We can take an equilateral triangle and cut along the angle bisectors, which meet at a point in the center.

(b) We can take an equilateral triangle and connect midpoints of its sides.

(c) Here we have a more complicated case. It can be done, for example, as shown in the diagram.

Problem 8.10. Let's see first what happens in the simpler case where nobody splits into two. In this case, after one day we have 1 Xork and 99 Yorks; the next day we have 1 Xork and 98 Yorks, and so on. Every day the number of Yorks goes down by one. Thus all Yorks will be destroyed in 100 days.

Let's get back to our problem, in which each day all Xorks and all Yorks split into two. Imagine the following thought experiment: after the first division, move half the Xorks and half the Yorks to a copy of the planet. So at the end of the first day there are two planets, each with 1 Xork and 99 Yorks. After another day and another division, move half the Xorks and Yorks from each of the planets to two more duplicate planets; there are now four planets, each with 1 Xork and 98 Yorks. We can continue in the same fashion, and we see that on any one planet the simpler case is playing out again, while the number of planets doubles each day. Therefore the Yorks will be destroyed after 100 days.

Problem 8.11. (a) Check that the example given on the right will work.

(b) Suppose, to the contrary, that we have placed 6 stars as required. Since the table has 4 rows and there are 6 stars, there must be a row which contains at least 2 stars. If it has at least 3 stars, erase this row and any other row containing a star. There will be 2 or fewer stars left, so we'll easily remove them by erasing at most 2 columns.

If the chosen row contains exactly 2 stars, then the remaining 3 rows contain 4 stars. Therefore at least two stars are contained in one row. Again, we can erase two rows that together contain at least 4 stars. The remaining stars can be removed by erasing at most 2 columns.

Solutions to Problem Set 9

Problem 9.1. Let's add 1 to the sum. We are now looking at the number $1 + 1 + 2 + 4 + 8 + \cdots + 1024$. Let's replace $1 + 1$ with 2. The result is then $2 + 2 + 4 + 8 + \cdots + 1024$. Replace $2 + 2$ with 4 to get $4 + 4 + 8 + \cdots + 1024$. We can continue replacing the sum of two equal powers of 2 with the next power of 2. Eventually we reach $1024 + 1024 = 2048$. The original sum was 1 less than 2048.

Does this problem remind you of the goose problem from the previous lesson?

Problem 9.2. Solution 1: First we need to decide if we are going to count meetings in port. Let's not. Now analyze which ships will meet a ship that leaves Belfast. There are 6 ships that are already heading to Belfast. One left 6 days ago, the next 5 days ago, and so on with the last having left 24 hours ago. In the next 7 days another 7 ships will leave Savannah and head for Belfast. One will leave at the same time as our ship, another 24 hours later, and so on. The last one will leave port 24 hours before our ship arrives in Savannah. Therefore we get a total of 13 ships. If we also count ships as they meet in harbor, there will be 2 more. One will be arriving in Belfast as the one we are tracking is leaving Belfast, and the other will be leaving Savannah just as our ship is arriving there.

Solution 2: Divide the distance between Belfast and Savannah into 7 different segments because the ship covers each segment in 24 hours. Ships leave port every 24 hours and all are traveling the same speed so they approach each other at double that speed. Therefore ships will meet every 12 hours and they will meet at the endpoints and midpoints of the segments. If we don't count the ports, we get 6 ends and 7 midpoints of the segments giving a total of 13 places where ships meet.

Solution 3: Let's say the ships that are heading to Belfast don't leave from Savannah but from a third city, let's say London, that is a 7 days' sail from Savannah. Think of Savannah as an intermediate point on the journey from London to Belfast. Because the ships leave London every 24 hours, when a ship leaves Belfast, there are 13 ships heading toward it not counting the one entering Belfast harbor or the one just leaving London. The Belfast to Savannah ship will meet these 13 ships on its voyage.

Set 9   99

Problem 9.3. Here we can use the triangle inequality: The sum of the lengths of any two sides of a triangle is greater than the length of the third side. Let us examine triangle ABC, where $AB = 3.8$ inches, and $BC = 0.6$ inches. Let us find the length of AC knowing that it is a whole number of inches. AC can't be 3 inches or less because the sum of the lengths of AC and BC would be $3 + 0.6$ which is less than 3.8, the length of AB. AC can't be 5 inches or greater because the sum of the lengths of AB and BC would be $3.8 + 0.6$ which is less than 5, a number at least as big as the length of AC. Therefore, there is only one option left, which is 4 inches.

There is, in fact, such a triangle. It can be constructed with a ruler and compass as follows. First construct leg AC of length 4 inches. Then draw a circle about A with radius 3.8 inches, and a circle about C with radius 0.6 inches. These circles will intersect. Either of the two points of intersection can be taken as vertex B.

Problem 9.4. The numerator is equal to $1 \cdot 2 \cdot 3(1 + 2 \cdot 2 \cdot 2 + 4 \cdot 4 \cdot 4 + 7 \cdot 7 \cdot 7)$, and the denominator is equal to $1 \cdot 3 \cdot 5(1 + 2 \cdot 2 \cdot 2 + 4 \cdot 4 \cdot 4 + 7 \cdot 7 \cdot 7)$. Thus the fraction is equal to $\frac{2}{5}$.

Problem 9.5.(a) The square depicted in diagram on the left consists of four half-cells, so its area is equal to the area of two cells.

(b) The square depicted in diagram on the right consists of a central cell and four adjacent right triangles. Each triangle has area equal to one cell because it is half of a rectangle formed by two cells. Therefore the total area of the square is equal to the area of five cells.

Problem 9.6. If $ABCD$ is our quadrilateral, then AC and BD are its diagonals. Using the triangle inequality, we conclude that $AB + BC > AC$ and $CD + DA > AC$. Adding these inequalities gives us that the sum of the sides of the quadrilateral must be greater than twice AC. Similar reasoning, which you can complete, shows that the sum of the sides of the quadrilateral is greater than twice BD. Since twice the sum of the sides of the quadrilateral is greater than twice the sum of its diagonals, the sum of the sides is greater than the sum of the diagonals.

Problem 9.7. We are assuming that the order in which George eats items at a meal doesn't matter; only which items he eats matters.

(a) We have already seen several problems like this one. While putting together his lunch, George can either choose or not choose each item. Therefore, he has a total of 2^{10} options, from not eating at all to stuffing himself by buying everything. So George can eat lunch 1024 times without repeating a meal.

(b) Let's construct a table where the rows correspond to days, and the columns correspond to items George buys. Place a "+" at the intersection of day n and item m if George eats that item on that day, and place a "−" if he does not. Therefore row n will consist of 10 signs which will indicate which items George bought that day and which ones he did not buy. There are 1024 rows in the table because that is the number of options for a meal, and there are 10 columns representing the 10 different items. To count the number of items George eats, we need to count the number of plus signs in the table. Because all of the different combinations of plus and minus signs occur once throughout the table, there are an equal number of each. The number of signs in the table is $1024 \times 10 = 10\,240$, and there are half as many plus signs, or 5120. George eats 5120 items over 1024 days.

We can also explain the fact that there are as many plus as minus signs in the table like this. Let's pair up one row of the table with another if each row is the result of changing every sign to the opposite. For example, $--+-+++---$ will be paired with $++-+---++$. All of the rows will match up to form such pairs, and between each pair there will be 10 pluses and 10 minuses. Therefore pluses form half of the table.

We can also explain it like this. Half of the signs in the first column are pluses and half are minuses because George eats the first item on the menu half of the days and does not eat it the other half of the days. The same occurs in all of the columns of the table. Therefore, half of the table is composed of pluses and half of minuses.

Problem 9.8. The pigeonhole principle will be helpful here again. There are 40 pairs of numbers between 10 and 100 that sum to 100. These are $(10, 90)$, $(11, 89)$, ..., $(49, 51)$. That leaves out the 10 two-digit numbers $50, 91, 92, \ldots, 99$. Even if we take all of these 10, and one number from each pair, we still end up with only 50 numbers. If we take more than 50 numbers we will have to take both numbers from at least one pair, and they will sum to 100.

Problem 9.9. Our star consists of a regular pentagon, which is the central part of the star, and 5 adjacent triangles, which we will call its rays. Draw two diagonals in the central pentagon, as shown on the right. They divide the pentagon into three parts: two small congruent triangles, and one slightly bigger triangle. The latter one is congruent to a ray of the star since it forms, together with the adjacent ray, a parallelogram. You should prove this.

Now we see that the shaded part of the star consists of three triangles congruent to a ray and one small triangle. The unshaded part of the star also consists of three rays and one small triangle. Since the two smaller triangles are congruent, the area of shaded region is exactly half of the total area.

Problem 9.10. Our sum is equal to

$$6 \cdot (1 + 11 + 111 + \cdots + \overbrace{11\ldots 11}^{100})$$
$$= \tfrac{6}{9} \cdot (9 + 99 + 999 + \cdots + \overbrace{99\ldots 99}^{100})$$
$$= \tfrac{2}{3} \cdot (10 - 1 + 100 - 1 + 1000 - 1 + \cdots + 1\overbrace{00\ldots 00}^{100} - 1)$$
$$= \tfrac{2}{3} \cdot (\overbrace{11\ldots 11}^{100} 0 - 100) = \tfrac{2}{3} \cdot (\overbrace{11\ldots 11}^{99} 00 - 90)$$
$$= 2 \cdot (\overbrace{37037\ldots 037}^{98} 00 - 30) = \overbrace{740740\ldots 740}^{99} 0 - 60 = \overbrace{740740\ldots 740}^{96} 7340.$$

Problem 9.11. Let's examine the numbers $1, 11, 111, 1111, \ldots$. There are infinitely many such numbers while there are 2003 possible remainders on dividing numbers by 2003. Therefore, we can find two numbers consisting only of 1's which have identical remainders when divided by 2003. The difference of these two numbers will be of the form $11\ldots 1100\ldots 00$, and will be divisible by 2003.

We can get an additional result from our conclusion. Because 2003 is not divisible by 2 or 5, the 0's at the end of the resulting number can be thrown away. The remaining string of 1's must be divisible by 2003, so there is a number that consists of only 1's that is divisible by 2003.

Problem 9.12. Suppose the conclusion is false, and we can't form a triangle from any three pieces. Arrange the pieces in ascending order of length. The two shortest ones are at least 17 cm long. Then the next one is at least 34 cm long because otherwise we could form a triangle from the three shortest pieces. Similarly, the next segment must be at least $17 + 34 = 51$ cm long, and the fifth must be at least $34 + 51 = 85$ cm long. Thus the five pieces together would form a bar at least $17+17+34+51+85 = 17(1+1+2+3+5) = 17 \cdot 12 = 204$ cm long. This is a contradiction, since 204 centimeters is greater than 2 meters.

Note: The numbers 1, 1, 2, 3, 5 found in solving this problem are the beginning of an infinite sequence called the Fibonacci numbers. The members of this sequence are formed by adding the previous two members. The Fibonacci sequence, $1, 1, 2, 3, 5, 8, 13, 21, \ldots$, appears often in mathematics.

Solutions to Problem Set 10

Problem 10.1. Let W stand for Winnie the Pooh's house, P for Piglet's and E for Eeyore's. Piglet covered a path of length $PE + EW + WP$, and Pooh covered a path of length $2EW$. Piglet's path is longer, because $PE + WP > EW$ by the triangle inequality.

Problem 10.2. Yes, it is possible. One way to achieve the goal is to fill the 3-liter beaker and transfer and liquid to the 5-liter beaker. Refill the 3-liter beaker and use it to fill the 5-liter beaker. Empty the 5-liter beaker and transfer the remaining 1 liter from the 3-liter beaker into the 5-liter beaker. Fill the 3-liter beaker and transfer its contents to the 5-liter beaker. Now the 5-liter beaker contains exactly 4 liters of water.

Problem 10.3. Let's give the first customer the remaining nut. Then the first customer has 2 nuts — the same number as the second. The first and second together will have 4 nuts — as much as the third customer. The first three together will have 8 nuts, the same number of nuts as the fourth customer, and so on. Therefore, the last customer will have as many nuts as all of the preceding customers put together. The nuts bought by the last customer weigh 50 lbs., so all of the nuts together weighed 100 lbs.

Problem 10.4. (a) Yes. Any even number of dollars can be paid in $2 bills, and any odd number greater than or equal to five can be paid with one $5 bill and several $2 bills.

(b) Yes. Notice first that 8, 9, and 10 tenges can be paid very easily: $8 = 3 + 5$, $9 = 3 + 3 + 3$, $10 = 5 + 5$. Any larger number can be reached easily by adding sufficiently many 3-tenge coins to 8, 9, or 10. For example, $11 = 8 + 3$, $12 = 9 + 3$, $13 = 10 + 3$, $14 = 8 + 3 + 3$, $15 = 9 + 3 + 3$, $16 = 10 + 3 + 3$, and so on.

Problem 10.5. (a) If the towns are on opposite sides of the road, the answer is direct. The stop should be located at the intersection of line AB and the road. Then the sum of the distances from A to the stop and B to the stop will be equal to the distance from A to B. For any other point on the road, the sum will be larger by the triangle inequality as shown in the diagram.

(b) A solution for this part of the problem is not as straightforward, but it turns out that we can reduce this problem to the previous case. Reflect point A symmetrically across the road so that we get point A'. Then for any point O on the road, the distance from A' to O is the same as the distance from A to O as one can see in the corner diagram.

Thus the sum of the distances from A and B to O is equal to the sum of the distances from A' and B to O. To solve this part of the problem we need only find a point O for which the sum of the distances from A' and B to O is the least. But we already solved this problem in (a) by connecting A' and B with a line segment and placing O at the point of intersection of that line and the road as shown in the diagram.

Problem 10.6. Because they walked the same number of minutes before they passed each other, Pooh got to Piglet's house three minutes before Piglet got to Pooh's house, so Pooh walked faster than Piglet. Let's say that Pooh walked x times as fast as Piglet. Then, before the meeting, the distance Pooh covered is x times greater than the distance Piglet covered. After the meeting Piglet covered a distance x times greater than Pooh did, and he did it at a speed of $\frac{1}{x}$ times Pooh's speed. Therefore, the time Piglet spent walking after the meeting is x^2 times greater than the time Pooh spent walking after the meeting. So we get that $x^2 = 4$, and thus $x = 2$. This says that Pooh is walking twice a fast as Piglet. Piglet took 4 minutes to go from the meeting point to Pooh's house, so Pooh took 2 minutes to cover the same distance. Therefore Pooh took $1 + 2 = 3$ minutes to go from his house to Piglet's. Piglet, walking at half the speed, took twice the time or 6 minutes.

Problem 10.7. Number the squares along the diagonal that runs from the lower left corner of the board to the upper right from 1 to 8. Square 1 belongs to some domino; for example in the diagram it is a vertical one. If the square to the right of 1 and adjacent to it also belongs to a vertical domino, we have found two dominoes forming a 2×2 square and the problem is solved. Otherwise, this square is part of a horizontal domino, and we end up with the picture depicted in the diagram. Now let's examine square 2. If it is part of a horizontal domino, then we have found a 2×2 square. Otherwise, square 2 is covered by a vertical domino. Then take square 3, and we continue as above. We gradually build a "Christmas tree" of dominoes and end up with a 2×2 square either before or when we reach the upper right corner of the board.

Problem 10.8. For a quadrilateral $ABCD$, let O be the point of intersection of diagonals AC and BD. Let us prove that O is the point we are looking for. To do so we take a point M in the quadrilateral with $M \neq O$ as in the diagram below, and prove that the sum of the distances from M to the vertices of the quadrilateral is greater than the sum of the distances from O to those vertices. By the triangle inequality $MA + MC \geq AC = OA + OC$, and the equality holds only when M lies on AC. Similarly, $MB + MD \geq BD = OB + OD$ with equality only when M lies on BD. Since $M \neq O$, at least one of these two inequalities is strict. Thus $MA + MB + MC + MD > OA + OB + OC + OD$.

Problem 10.9. (a) In this case, there is a solution you may already have heard of. One of the pirates divides the treasure into two parts that are equal from his perspective, and the second chooses a part that he thinks is at least as great as the other part. In this way both are satisfied they have a fair share.

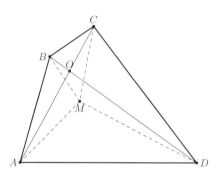

(b) First, two pirates should divide the treasure among themselves so that each thinks he got at least half. This can be done as indicated in (a). Then the third pirate divides each of the parts into three parts that are equal from his perspective. The first pirate then takes the two "best" parts of the three his half has been divided into. The second pirate does the same with his half. The third gets the remaining thirds from each of the two original piles and ends up with one third of the total as he perceives it.

(c) We can proceed similarly as in (b). First, three pirates divide the treasure among themselves and then the fourth takes a fourth from each portion. We think you can fill in the details of the process.

(d) Solution 1: We can proceed as in (c), consecutively explaining how to divide the treasure into 4, 5, 6, or more parts.

Solution 2: With any number of pirates, we present a process to first give one pirate his portion in such a way that he will be happy and the others will not object. This reduces the problem to a smaller number of pirates. We continue this process until everyone has obtained his part.

For ease of explanation we will assume that there are ten pirates, but you can easily change the details for any number of pirates. The pirates stand in one line. The first will take what is, in his opinion, $\frac{1}{10}$ of the treasure. Then he walks past the line of pirates with this part and asks each in turn the same question: "Do you agree that I took no more than $\frac{1}{10}$ of the treasure for myself?" If all of them agree, he takes this part and leaves. If one of them, let's say the third one, does not agree, the first will tell him: "Then you should return the excess to the pile so that only one-tenth remains, and take that tenth." The third pirate must do so, after which the third will continue along the line from the fourth to the tenth, asking the same question. If all of them agree, then the third one leaves with his part.

Set 10   105

Otherwise they repeat the above process. Eventually, one of the pirates will be able to leave, having gathered what he considers $\frac{1}{10}$ of the treasure. Everyone is now sure that he took no more than $\frac{1}{10}$ of the treasure so at least $\frac{9}{10}$ remains. The problem will be solved, if the remaining 9 pirates can divide the treasure in such a way that each will get no more than $\frac{1}{9}$ of the treasure. But the remaining 9 can act just like the 10 did before. Then one of these 9 will leave with his part, and so on, until all of the pirates will have their part of the haul.

Problem 10.10. This is similar to Problem 7.11 and can be approached in a similar manner. Construct a circle of radius 2 meters with its center at one of the vertices of the square and color all of the points inside the square that lay on this circle or outside it. We get a curvilinear triangle as drawn in the diagram on the left. Two sides of the curvilinear triangle are sides of the

 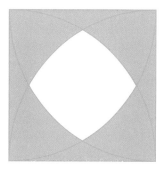

square and the third is one-fourth of the constructed circle. Do the same for the four remaining vertices of the square. We end up with the green region in the diagram on the right. The grasshopper can reach any points in the green region of this figure, and no other points. Try to prove this yourself using some ideas from the proof in Problem 7.11.

Problem 10.11. Before cutting our 11×11 grid, let us mark sixteen 2×2 squares as indicated in the diagram. Notice that no matter how we cut a 2×2 square along the gridlines, it can't contain grid squares from more than one of the marked squares. Because we cut out fifteen squares and there are sixteen marked squares, we can find at least one marked square that has no grid squares in common with any of the squares we cut out. Then we can cut out that marked square.

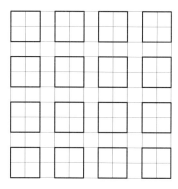

Solutions to Problem Set 11

Problem 11.1. No. We can take any rhombus that isn't a square, like the one in the diagram. When folding it along any diagonal its edges will coincide.

Problem 11.2. Solution 1: Let's first examine the simplest case of the first four numbers: 1, 2, 3, and 4. Their sum is equal to 10 and is not divisible by 4. What will the sum of 2, 3, 4 and 5, the next four numbers, be? Each number increases by 1, so the sum will increase by 4, and again it will not be divisible by 4. We can repeat this argument and show the sum of four consecutive numbers is never divisible by 4.

Solution 2: Take four consecutive natural numbers. Let the smallest be n. Then the rest will be $n+1, n+2$, and $n+3$. Their sum is equal to $4n+6$, which is not divisible by 4 because $4n$ is divisible by 4 and 6 is not.

Problem 11.3. (a) The second player can guarantee victory. The second player should take the same number of stones as the first player did on the previous turn, but from the other pile. The second player is making moves "symmetrical" to those of the first. Then the second will always be able to make the last move, and will always win.

(b) The first player wins by taking one whole pile and can now symmetrically answer the second player's moves.

(c) The second player can win by dividing the piles mentally into two pairs. Then if the first player takes stones from a pile of one of these pairs the second takes the same number of stones from the second pile of this pair. In other words, in each pair of piles, the second player moves symmetrically to the first player.

Problem 11.4. If we rip one piece of paper into four pieces, the total number of pieces increases by 3 — the one we ripped up was removed and replaced with four new pieces. If we rip one piece into six pieces, the total number of pieces will increase by 5. Therefore we can reformulate our problem as follows. First we have 1 piece. During each turn, we can increase the number of pieces by 3 or 5. Prove that, using a series of such moves, we can attain any number of pieces from 9 on up. Now the problem looks like Problem 10.4(b).

Set 11

Problem 11.5. Very often people think that the company will cut down 1% of the forest, but this is incorrect. Let's look into the situation more carefully. The trees that aren't pines make up 1% of the original forest. After cutting down some of the pines, the number of trees that aren't pines will not change, but they will now make up 2% of the total. This is possible only when the trees that are not pines together with the remaining pines are half the size of the original forest. The company proposes to cut down half of the forest.

Problem 11.6. This problem's solution is very similar to that of Problem 11.4. If we already have cut the original square into several squares, we can now cut one of these into four smaller squares by connecting the midpoints of its opposite sides. This will increase the total number of squares by 3. If we can find out how to cut one square into 6, 7, or 8 squares then, by starting with one of these numbers and increasing the number by 3 as many times as we need to, we can cut one square into 6 or more parts. The diagrams above show how to cut a square into 6, 7, or 8 smaller squares.

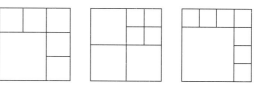

Problem 11.7. Solution 1: Let's first solve the case when A is on the bank of the canal. To get to B we need to walk some distance along A's bank to some point C, cross a perpendicular bridge to a point C' on the other bank and finally walk directly to point B. The length of the bridge will be the same no matter where it is built, and it doesn't matter if we first walk along the bank and then cross or first cross and then walk along the bank. Thus we can first cross to the other side, arriving at point A' on the bank directly opposite A. The problem has now been reduced to the following. We need to find some point C' on the bank with B for which the sum of the distances $A'C' + C'B$ is the least. Then we should build the bridge at this point. The straight line from A' to B is the shortest path between them so C' and A' will coincide. See diagram.

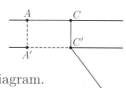

We now generalize the solution to the case when A is not right on the bank. In the case above, we introduced a new point A' that is on a line from A perpendicular to the canal and is the width of the canal closer to B's side. Let's do this again and consider a point A' obtained by translating A toward the canal by the length of a bridge in a direction perpendicular to the canal. Then, if CC' is the bridge, with C on the same side of the canal as A, the sum of the distances $AC + CC' + C'B$ will be equal to the sum of the distances $AA' + A'C' + C'B$ because the polygon $AA'C'C$ is a parallelogram, as shown on the right.

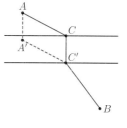

The length AA' is the width of the canal and does not depend on the position of the bridge, while the sum $A'C' + C'B$ will be minimized if it is equal to $A'B$. Therefore we should let C' be the point where the line joining A' and B intersects the bank closer to B. One needs to recognize that A' is not a point on the road, but is a point on the diagram. It helps us solve the problem, and it does not matter if it lands in the canal.

Solution 2: Let's try to solve this by analogy with Problem 10.5(b). Let CC' be the bridge with C on the same side of the canal as A, and C' on the same side as B. Reflect the side with A across the line down the middle of the canal as in the diagram. The point A reflects to A' and C reflects to C'. Introduce one more point, E, the midpoint of the bridge. The length of the trip from A to B is $AC + CC' + C'B$, but this 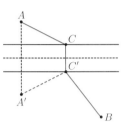 equals $A'C' + 2C'E + C'B$. We want to make this as small as possible, and $2C'E$ is the width of the canal which is fixed so we need to choose C' so that $A'C' + C'B$ is as small as possible. This is Problem 10.5(b), which we already know how to solve.

Problem 11.8. Draw a gridded rectangle corresponding to the pool table with its longer side horizontal and its shorter side vertical. Assume the ball is shot from the lower left corner.

(a) Solution 1: The diagram makes it easier to follow the path of the ball. In the diagram we see that the ball will rebound six times, before reaching the upper right corner.

Solution 2: This solution will at first look complex, but it will help us solve part (b) and many similar problems involving billiards (pool).

Let $ABCD$ be the rectangle representing the pool table, and shoot the ball from point A. After each collision of the ball with a side of the table we will symmetrically mirror the rectangle across this edge and draw the continuing path of the ball in the resulting rectangle. Because the angle of incidence is equal to the angle of reflection, the trajectory of the ball becomes a straight line, as we see in this sequence of diagrams:

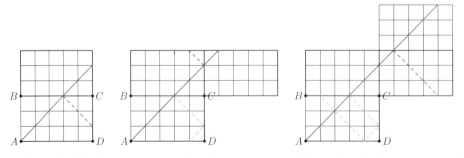

Now divide the plane into 3×5 rectangles. We will call the vertices of these rectangles nodes, and the lines of their sides horizontals and verticals. Draw a ray from A at a 45-degree angle to the sides of the pool table. Then the next node the ray passes through will correspond to the first corner the ball reaches, and the number of intersections of the ray with the horizontals and verticals before reaching the node will correspond to the number of collisions the ball makes with the sides of the table. Let's carry out the details.

Let X be the next node the ray passes through after A. Since the ray moves along the bisector of angle DAB, the segment AX is the diagonal of a square whose vertices are grid nodes. Because the distance between neighboring horizontals is 3, this square will have side divisible by 3. Because the distance between neighboring verticals is 5, the square will have side divisible by 15. The smallest natural number that is divisible by both 3 and 5 is 15, so the square is 15×15. Its diagonal 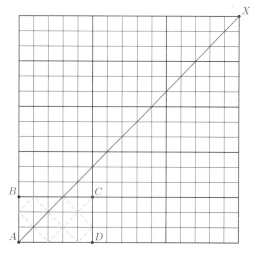 crosses two verticals and four horizontals within the square, meaning that on the diagonal there are 6 points of intersection between the diagonal and sides of rectangles other than A and X. These points correspond to the six collisions of the ball with the sides of the pool table. For the diagonal AX, the ball starts at A, and X represents the corner where the ball ends up and is illustrated in the figure above. How can we find out which vertex of the original rectangle X represents?

To do so, every time when we reflect a rectangle, we will indicate at each node which vertex of the original rectangle falls on this node. We end up with the picture depicted in the diagram. We can see that along each horizontal letters alternate, and likewise letters alternate along each vertical. Along the first, third, fifth, and every odd horizontal the alternating letters are A and D, while the letters alternating along the second, fourth, and every even horizontal, are B and C. Similarly, 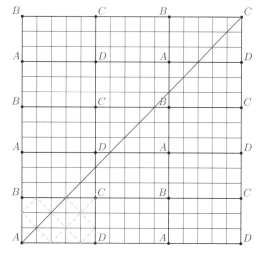 along the first, third, fifth, and every odd vertical the alternating letters are A and B, while along the second, fourth, and every even vertical D and C alternate. Therefore, we can figure out which letter is at what node. A is always at the intersection of an odd horizontal and an odd vertical, B is at the intersection of an even horizontal and an odd vertical, C is at the intersection of an even horizontal and an even vertical, and D is at the intersection of an odd horizontal and an even vertical. Since X is at the intersection of the sixth horizontal and fourth vertical, it corresponds to letter C. Thus the ball will end up at corner C of our pool table.

(b) Solution 1: Let's try to follow the path of the ball while counting the number of times it collides with the sides of the table. To reach the first collision the ball will move 3 cells horizontally. It will move another 3 cells to reach the second collision, and so on. Having gotten to the 333rd collision, the ball will have moved 999 cells horizontally. Because collisions alternate between the upper and lower sides with odd collisions on the upper side and even ones on the lower, on the 333rd collision the ball will be bouncing off the upper side and will have two horizontal units to travel before reaching the right side. The next collision will be on the right vertical side 2 units down. The 335th collision will be on the bottom side 1 unit from the right side as shown here:

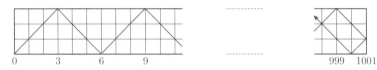

After this the ball will have another 333 collisions with the horizontal sides before colliding with the left vertical wall 1 unit down and then hitting the bottom side 2 units from the left side:

After this the ball will have 332 collisions with the horizontal sides and will finally reach the upper right corner. The total number of collisions is $333 + 2 + 333 + 2 + 332 = 1002$.

Solution 2: Let's try the approach used in Solution 2 for (a). Divide the plane into rectangles $ABCD$ of size 3×1001, and send a ray from A at a 45-degree angle to the sides of the rectangle. Let X be the next node the ray passes through. Segment AX will be the diagonal of a square with vertices at nodes. A side of the square must be divisible by 3 and by 1001, and therefore is divisible by 3003. Therefore our square is 3003×3003. Its diagonal crosses two verticals and 1000 horizontals within the square, so on the diagonal AX there are 1002 points of intersection between the diagonal and sides of rectangles other than A and X. These points correspond to 1002 collisions of the ball with the sides of the pool table. Since the node X is at the intersection of the 1002nd horizontal and the fourth vertical, it corresponds to point C of the original rectangle, and therefore the ball ends up at the right top corner of the table.

Set 11

Problem 11.9. Yes, it is always possible. Let's say we have a sheet divided into polygons by several lines. First erase these lines, and then put them back one by one coloring the plane as we add lines so that the color scheme of the plane satisfies the conditions of the problem. Assuming the sheet of paper is white, draw the first line and color the polygon on one side of the line black. We get a coloring which satisfies the requirements. At each successive step we do the following: Draw a new line, pick one side of the line where the colors will not change but reverse all the colors on the other side of the line switching white to black and black to white. On opposite sides of any old line, neighboring areas will be colored differently because they were different colors before we drew the latest line and either their colors were unchanged or reversed. Any two polygons that are separated by our new line will also be of different colors because, before we drew the new line, they were just one polygon which we cut and changed the color on one side. Thus we consecutively draw all of the lines, and we get a color scheme in which polygons that share a side are different colors.

Problem 11.10. (a) No. For example, let's take nine rods of length 1 inch and one of length 100 inches. We can't form a decagon from these rods.

(b) Not always. For example, suppose that one quadrilateral has sides of lengths 2.5, 1, 1, and 1, and the other two are squares with side 1. In this case it is impossible to build four triangles using these 12 rods. Indeed, we have one rod of length 2.5 and all other rods of length 1. One of the triangles would have to contain one long rod and two short ones, but these three rods can't form a triangle since the sum of the lengths of two short rods is less than the length of the long rod.

(c) Yes. Let us prove this. But how can one prove that a polygon can be built using rods of given lengths? The following criterion is intuitively true.

It is possible to build a polygon from a collection of rods provided that the length of the longest rod in the collection is less than the sum of the lengths of all other rods.

Using this criterion, we show how to construct three quadrilaterals from the four triangles. For each triangle, find the length of its longest rod. Choose a triangle for which this length is least. In other words, choose a triangle whose longest side is no longer than the longest side of the other three triangles. We distribute the rods of this triangle among the remaining triangles by adding one of its rods to each of the other triangles; by our choice of triangle, no rod distributed to another triangle becomes the longest. Thus the longest rod in each quadrilateral is the same as the longest rod in the triangle the quadrilateral was made from. In each triangle, the longest rod has length less than the sum of the lengths of the other rods so adding an additional shorter rod does not change this fact. By the criterion, we can form three quadrilaterals if we distribute the rods as stated.

How can we be certain that the criterion, which seems obvious, is indeed true? For triangles, a proof is not too hard.

If a triangle can be built, the length of the longest side must be less than the sum of the other two sides by the triangle inequality.

But why is the converse true? If the length of the longest rod is less than the sum of the lengths of the other two rods, why it is always possible to build a triangle? Let us draw a line segment AB representing the longest rod and construct two circles: one centered at A with radius equal to the length of the second rod, 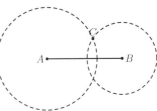 and the second circle centered at B with the radius equal to the length of the third rod. Both these circles will intersect AB since the length of AB is the largest. Moreover, the point of intersection of the first circle with AB will be closer to B than the point of intersection of the second circle with AB since the sum of the radii of the circles is bigger than the length of AB. This means that the circles will intersect as we see above. The point of intersection C will provide the third vertex of the desired triangle ABC.

Let's think about how we might construct a polygon with n sides from n rods if the longest has length less than the sum of the lengths of the other rods. Take a very large circle, and place our rods one after another so that they form a broken line inscribed into some arc of the circle:

Let the longest rod be the first segment of this chain, and imagine that the segments are hinged at the vertices and they can pivot on the hinges but cannot leave the circle. As we make the circle smaller, the chain will

Set 11 113

remain inscribed while its beginning and ending points will move along the circle closer and closer to each other. At what moment will we be unable to continue squeezing the circle? Either the beginning and endpoints will merge and the broken line closes up to form the desired polygon, or the longest rod becomes a diameter of the circle. In the latter case, the longest rod splits the circle into two arcs, one of which contains the other rods. If this happens start expanding the circle by pulling its center away from the string of shorter rods while keeping the ends of the longest rod on the circle. The arc with the rest of the rods will get closer and closer to the longest rod, and, because the sum of the lengths of these rods is bigger than the length of the longest rod, the endpoints will move closer and at some moment the chain will become closed, giving us our polygon.

Problem 11.11. At first we can try to guess the answer. We can figure out that $16 = 4^2$, $1156 = 34^2$, $111556 = 334^2$, and we can hypothesize that the numbers are squares of the numbers $4, 34, 334, 3334, \ldots$. Let us check this. Suppose we take the n-digit number $33\ldots34$ and square it. We have

$$\underbrace{33\ldots4}_{n \text{ digits}}{}^2 = (\underbrace{33\ldots3}_{n \text{ digits}}+1)^2 = \left(\tfrac{1}{3}\cdot(\underbrace{99\ldots9}_{n \text{ digits}}+3)\right)^2 = \tfrac{1}{9}(10^n+2)^2$$

$$= \tfrac{1}{9}(10^{2n}+4\cdot 10^n+4) = \tfrac{1}{9}(\underbrace{99\ldots9}_{2n \text{ digits}}+1+4\cdot(\underbrace{99\ldots9}_{n \text{ digits}}+1)+4)$$

$$= \underbrace{11\ldots1}_{2n \text{ digits}}+4\cdot\underbrace{11\ldots1}_{n \text{ digits}}+1 = \underbrace{11\ldots1}_{n \text{ digits}}\underbrace{5\ldots56}_{n \text{ digits}}.$$

Solutions to Problem Set 12

Problem 12.1. The least number of lines is 1, when all 20 points lie on one line. The greatest number will be when all of the lines are different. This can occur, for example, when 20 points have been placed around a circle. Then no three points will lie on one line because any line crosses the circle in at most 2 points. With no three points on the same line, there are $\frac{20 \times 19}{2} = 190$ lines. This is exactly as many as the number of distinct pairs of points chosen from 20 points when the order of choosing does not matter.

Problem 12.2. The last page of the section has an even page number, so it ends in either 4 or 6. In addition, the number of the last page is greater than 463, so it starts with 6. Thus the only possible option for the last page is 634, so there are $634 - 462 = 172$ pages in the section. This means that there are $172/2 = 86$ sheets of paper in the section.

Problem 12.3. This is very similar to Problem 11.4. By ripping up one of the pieces, Mike increases the total number of pieces by 7. Therefore he can end up with 1+7 pieces, 1+7+7 pieces, and so on. Subtracting 1 from the number of pieces always leaves a number that is divisible by 7, but 2001 is not divisible by 7. Therefore Mike can't rip the paper into 2002 pieces, but he could rip it into 2003 pieces.

Problem 12.4. The second player can win. After each turn by the first player, the second player returns the marker to the diagonal that leads from the lower left corner of the board to the upper right corner. The second player replies to any vertical move by the first player with a similar horizontal move, and to any horizontal move by the first player, with a similar vertical move. (Compare with part (a) of Problem 11.3.)

Problem 12.5. As in Problem 11.6, it is enough to provide examples of cutting an equilateral triangle into 6, 7, or 8 equilateral triangles. This is because we can increase the number of triangles by 3 by cutting along the lines joining the midpoints of the sides of one triangle, and we can do this as often as we wish. Here are examples of how to make 6, 7, and 8 triangles:

Set 12   115

Problem 12.6. Let Alex take steps of length x inches at the rate of y steps per minute. Then Sam takes steps of length $1.1x$ inches at the rate of $0.9y$ steps per minute. In one minute Alex covers xy inches, while Sam covers $(1.1x)(0.9y) = 0.99xy$ inches. Sam is walking more slowly, and will arrive at school after Alex.

Problem 12.7. (a) For this part, one can draw the path of the ray while calculating each angle, as shown on the right. The third reflection will be made at an angle of 90 degrees to the mirror, after which the ray will return along the same path and will undergo a total of five reflections.

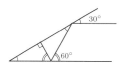

Drawing the path of the ray worked in (a), but when the angle between the mirrors is small, it is difficult to draw the path of the light ray to solve the problem. One can approach the problem in an alternative way. Each time the ray hits a mirror we will reflect the *wedge* between the mirrors in the mirror hit by the ray. We think of this as "unfolding" the region in which the ray travels. Because the angle of incidence is equal to the angle of reflection, in this unfolding the ray is no longer getting reflected but continues propagating in the same direction as it goes "through" the mirror. This is exemplified here for the two reflections shown in the previous diagram:

Each reflection of the ray corresponds, in the unfolded view, to an intersection of the straight trajectory of the ray with the sides of reflected angles. For a 30° wedge it takes five reflections to unfold the region the ray travels in all the way to 180°, at which point the ray 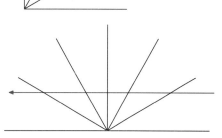 no longer hits mirrors. Hence the ray gets reflected exactly five times in the 30°-angled mirrors.

(b) Following the same unfolding method, we see that the ray undergoes eight reflections before exiting:

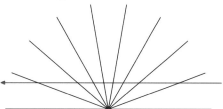

(c) In this case the ray is reflected only three times before exiting (see figure).

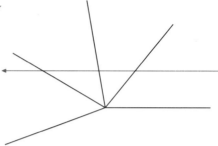

Problem 12.8. Solution 1: This solution is based on the fact that the sum of two identical powers of 2 can be replaced with a single power of 2: $2^n + 2^n = 2 \cdot 2^n = 2^{n+1}$.

Let us write our number N as the sum of N 1's. By combining these 1's in pairs we will get either a sum of 2's or a single 1 plus a sum of 2's. If we have two or more 2's, combine them in pairs to get a sum of 4's, at most one 2, and at most one 1. If we have two or more 4's, combine them in pairs to get a number of 8's, at most one 4, at most one 2, and at most one 1. Continuing in the same fashion, we'll obtain a representation of N as the sum of different powers of 2

Solution 2: Suppose we have a number N. Take the largest power, 2^k, of 2 not exceeding N. If $2^k = N$, stop because we have the desired representation. Otherwise subtract it from N, getting $N - 2^k$, and continue working with this new number in the same way. Find the largest power, 2^l, of 2 which does not exceed $N - 2^k$, and subtract it from $N - 2^k$. If $N - 2^k - 2^l = 0$, stop and N is expressed as the sum of two different powers of 2. Otherwise subtract the highest power of 2 not exceeding $N - 2^k - 2^l$ and continue in this manner. Since the numbers which we obtain at each step are getting smaller and smaller, eventually we will reach 0, and will get a representation of N as the sum of powers of 2. All these powers will be distinct because, if we were able to subtract the same power of 2 twice, we should have subtracted a larger power of 2 to begin with.

We have shown every natural number can be expressed as a sum of distinct powers of 2. Is it possible that some natural number can be expressed in two different ways as a sum of powers of 2? We can always rearrange the summands as in $7 = 4 + 2 + 1 = 2 + 1 + 4$, but could we have different exponents?

Suppose there are two different ways to represent the same number as the sum of powers of 2. Equate these two sums, and cancel those powers of 2 which occur on both sides of the equation. We will get an equation where all terms on the left have different exponents from all terms on the right. Now let us find the smallest number used in the equation. This will be smallest power of 2 occurring in the equation, and hence every other number in the equation will be divisible by it. Let us divide both sides of the equation by this smallest power of 2. The term we divided by will become 1, while every other term will be even. Thus we will get an equation having an even number on one side and an odd number on the other side. This is a contradiction, so we could not have expressed one number as a sum of different powers of 2.

Set 12   117

Problem 12.9. The second row of our table holds the numbers $4+1, 4+2, 4+3, 4+4$. These are the numbers of the first row with 4 added to each. If we place a plus before two of the numbers and minus before the other two, two of the added 4's will become positive, and two negative, and the additional 4's will total 0 when we sum the second row. Consequently summing the second row with signs inserted gives the same result as summing a row with $1, 2, 3, 4$ and the corresponding signs we inserted in the second row. Therefore, for putting in signs and summing the entries, we can assume that the second row in the original table was the same as the first. Similarly, we can say that the third and fourth lines are $1, 2, 3, 4$, the same as the first. But then we have a table with four 1's in the first column, four 2's in the next, and so on. Inserting the signs puts two pluses and two minuses in each column so the sum of each column will be 0. Therefore, the sum of all of the numbers in the table will be 0.

Problem 12.10. Yes, it is always possible. Here is a way of showing it. Take a line somewhere far away from our points so that all of the points are to one side of it. Now move the line parallel to itself towards the points. With this movement of the line, the points "jump" across the line. If they "jump" across one at a time, then we can halt the line's movement when half of the points are on one side.

What if two or more points "jump" over the line simultaneously? This can only occur when the line we are moving is parallel to a line that connects those points. Let's draw red lines through every pair of points. There will be finitely many red lines, but there are infinitely many directions on the plane. We can start with a line that is not parallel to any of the red lines. When moving this line parallel to itself, points will "jump" over one at a time, and we can stop when half are on each side.

Problem 12.11. Let's start by considering an example. If we take the numbers $1, 2, 3, 4, 5, 6, 7, 8, 9, 10$, their sum is 55. We can see that these numbers do not work because $2 \cdot 3$ is not divisible by 55. However, we can adjust these numbers so that the new numbers satisfy the conditions of the problem. Specifically, multiply each number by 55, the original sum. We obtain the numbers $1 \cdot 55, 2 \cdot 55, \ldots, 10 \cdot 55$. Now the sum of these numbers is $1 \cdot 55 + 2 \cdot 55 + \cdots + 3 \cdot 55 = 55 \cdot 55$. The product of any of the two new numbers is the product of the two corresponding old numbers multiplied by 55 twice. The product of any two is divisible by the sum of all of them.

Solutions to Problem Set 13

Problem 13.1. This will be a circle with its center at the stake and radius equal to the length of the rope.

Problem 13.2. Draw circles of radius 1 meter centered on each of the vertices of the rectangle. Connect adjacent circles with line segments that are tangent to the circles, end at the points of tangency, are parallel to the sides of the rectangle, and lie outside the rectangle. The
"rectangle" with rounded corners that you have drawn is the outer limit of the goat's grazing. Now draw line segments that are tangent to adjacent circles, parallel to the rectangle's sides, but inside the original rectangle. These cut out a 1×3 rectangle inside the 3×5 rectangle. This is the inner limit of the goat's grazing. The diagram shows that the goat can consume all the grass between these two shapes.

Problem 13.3. The indicated region is bounded by two arcs of different circles. Let O_1 and O_2 be the centers of these circles, and let r_1 and r_2 be their radii. Attach the goat to O_1 with a rope of length r_1, and to O_2 with rope of length r_2. The first rope will not allow the goat to leave the area bounded by the first circle, while the second will not let it leave the area bounded by the second. The goat will be able to reach all of the points in the area of intersection of the two circles.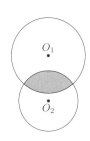

Problem 13.4. When the goat is tied to a specific point, it eats a circle about that point. Here the goat can eat a circle of radius the length of the second rope about any point that lies on the first
rope. By shading all such circles, we end up with the shape in the diagram, seen in many stadiums and racetracks.

Problem 13.5. Let's say we know how to tie the goat so that it can eat only the grass in figure F, and we know how to tie the goat so that it can only eat the grass in figure G. Then we can tie the goat so that it can only eat the grass in the intersection of F and G by tying it in both ways. Since we know how to keep the goat in any circle or stadium shape (Problem 13.4), this solves all of the parts of this problem, as follows:

(a) A semicircle is the intersection of a circle and a stadium shape, as shown in the leftmost figure below.

(b) A square is the intersection of two identical stadium shapes, placed at right angles to one another.

(c) A rectangle is the intersection of two distinct stadium shapes, also placed at right angles to one another.

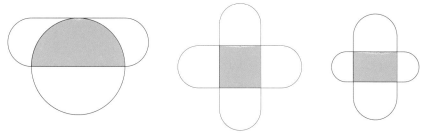

Problem 13.6. This problem can be solved using the previous observation, because a triangle can be represented as the intersection of three rectangles, and a regular hexagon can be represented as the intersection of two triangles:

Problem 13.7. (a) We can tie the dog to the same stake as the goat, with the dog's rope shorter than the goat's. See the leftmost diagram below.

(b) If the goat is tied to a rope staked in the ground, and the dog is tied to stay in a stadium shape with a straight side on a diameter of the goat's circle, the goat will stay in the semicircle that is inside the circle but outside the stadium.

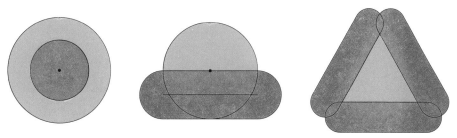

Problem 13.8. (a) Yes. The answer is based on the following fact. The circle circumscribed about a right triangle has center the midpoint of the hypotenuse, and radius equal to half the length of the hypotenuse.

Let ABC be our triangle. Tie goats to the centers of AB and BC. Drop height BH from vertex B onto leg AC. Triangle ABH is a right triangle, so the center of a circle circumscribed about it will be the midpoint of AB, where we have tied one goat, while the radius of that circle will be the length of the rope on the goat. Thus all of ABH is covered by the circle that one goat can eat. Similarly, all of CBH is covered by the other goat's circle. ABC consists of ABH and CBH when H drops onto AC, or is contained within one of these when H drops onto the continuation of AC. Therefore, the goats will be able to eat all of the grass in ABC.

(b) Yes. Divide the quadrilateral diagonally into two triangles. Then, according to (a), two goats can eat the grass in one of these triangles, and the other two can do so in the second. Therefore the four goats together can eat all of the grass in the quadrilateral.

Problem 13.9. Yes. If Peter was born on December 31 and today is January 1 of the year in which Peter will turn 12, then he turned 11 on December 31 last year. Then on December 30, the day before yesterday, he was 10. Next year Peter will turn 13.

Problem 13.10. At first, it's hard to see where to start with this problem. One way is to start at the shortest tree and walk around all of the trees in ascending order of height. Then we will walk no further than the difference in heights between the tallest and shortest trees or less than 25 meters. You may have to think about why this is so. You can return the same way or go directly from the tallest tree to the shortest and again cover at most 25 meters. Can you figure out a round trip path of no more than 50 meters if you don't get to pick which tree to start from?

Problem 13.11. The solution of this conundrum found by the Collector of Offerings was remarkably simple and ingenious. He placed the hippo into the empty boat, and marked on its hull the level to which the boat has submerged in the river. He then removed the hippo and began to load the boat with gold ingots. Once he has loaded enough ingots for the mark on the boat to once again reach the water, he truthfully reported to the Chief that the weight of gold in the boat was exactly equal to the weight of the hippo.

Solutions to the Winter Competition

Problem W.1. The five times the mailman collects mail from the mailbox split the day into four separate intervals. By dividing 12×4, we find that the mailman walks up to the box once every 3 hours.

Problem W.2. No, not necessarily. Cut a regular pentagon from a piece of paper. Draw a line segment parallel to and near one of its sides and cut a piece of the pentagon off along this segment as in the diagram. The resulting pentagon will have five equal angles, but it will not be a regular pentagon because its sides will not be equal.

Problem W.3. We need to fry a total of six sides and can fry at most two sides simultaneously so we will have to spend at least three minutes. Also three minutes will suffice because we can start by frying two pieces on one side for one minute. Then flip the first one, and replace the second with a third. After another minute, remove the finished first piece and return the second one in order to finish frying its unfried side. At the same time we flip the third piece over. After one minute more all three pieces of bread will be fried on both sides.

Problem W.4. Most likely there are three black balls and one white one in the box. There are at least two black balls in the box because two get pulled out. All of the balls cannot be black because we would always remove two black balls. Therefore, there are either two or three black balls. If there were two black balls, there would be two white balls, and we would expect to remove two white balls approximately 50 times also. Thus we would almost never remove a white ball and a black ball, which is unlikely. Since two black balls is very unlikely, the most likely is three black balls in the box.

Assuming there are three black balls and one white one, we can see how likely it is that we would pull out two black balls. Two black balls can be chosen from three black balls three different ways. A white and black ball can also be chosen three different ways from three black and one white.

Therefore, if we pull the balls out randomly, about half the time we would pull out two black balls, and half the time one black and one white.

Problem W.5. Yes, but only if she plays or does homework while in school. If the statement of the problem is interpreted as saying that none of these activities overlap, the hours don't add up: $\frac{1}{3}+\frac{1}{4}+\frac{1}{5}+\frac{1}{6}+\frac{1}{7}=\frac{153}{140}>1$. However, if Alice can (for instance) do math homework at school, it may be that the excess time of $\frac{13}{140}$ is simply being counted twice. For instance, Alice's day could go like this:

$\frac{11}{70}$	school, apart from math homework
$\frac{13}{140}$	math homework in school
$\frac{1}{20}$	math homework outside of school
$\frac{1}{5}$	volleyball
$\frac{1}{6}$	video games
$\frac{1}{3}$	all the rest
1	total

Problem W.6. This solution is based on the fact that every straight line passing through the center of a rectangle splits the rectangle into two parts having equal areas. Using the pencil and straightedge, we draw diagonals of the big rectangle, thus finding its center at the intersection. Do the same for the small rectangle. The line going through these two centers divides each of the two rectangles in half, and therefore divides the shaded region, as desired.

Problem W.7. Let $a_1, a_2, \ldots, a_{1000}$ be our numbers. Consider new numbers $a_1, a_1+a_2, a_1+a_2+a_3, \ldots, a_1+a_2+\cdots+a_{1000}$. There are 1000 new numbers, and 1000 remainders upon division by 1000. If the new numbers have different remainders, then among them we will find remainder 0, so the sum of several old numbers, or just the number a_1 will be divisible by 1000. If none of the remainders is 0, then we will find two new numbers with identical remainders. Their difference will have remainder 0, and be divisible by 1000. Subtract the one with fewer terms from the other, and we will find a sum of several consecutive old numbers, or a single old number, divisible by 1000.

Problem W.8. We need to find the sum of the angles A, B, C, D, E in the diagram. We will solve this problem with the help of a sewing needle. The solution is clearer if the needle is shorter than each of the sides of the star.

Run the following experiment. Place the needle along the side AB with the eye at point A. Slide it along AB until its point is at B, and turn it counterclockwise through the angle B so that it is now aligned with side BC. Continue by sliding the needle along BC until its eye is at point C, turn it counterclockwise around C through the angle C so that the needle aligns with side CD, slide it along CD until its point is at D, and so on. Finally, when the needle has slid along side EA to point A, turn it about A counterclockwise through the angle A. What has happened? If you watched the needle — real or imaginary — carefully, you would have noticed that in the process the needle has reversed its direction having turned by 180 degrees. On the other hand, we have turned it exactly trough the sum of the angles of the star. Therefore, the sum of these angles equals 180 degrees.

This method can be used to show that the sum of all exterior angles of any convex polygon is 360 degrees. Let's imagine that someone starts strolling along the sides of a convex polygon. The person starts at a vertex, walks along a side to the next vertex, turns at that second vertex, continues walking along the second side, and so on. On reaching the original vertex and turning, the person is at the original position. As a result, the person has made a complete circle and turned through the angle of 360 degrees. On the other hand, at every vertex, he was turning through the angle equal to the exterior angle at that vertex.

Problem W.9. (a) We can depict the problem as a drawing. Place 20 red dots in a row to represent the problems. Underneath them, draw several blue dots representing students. We also have a set of line segments to draw. A red dot is connected to a blue dot if the student represented by the blue dot solved the problem represented by the red dot. We know that two line segments leave each red dot because each problem was solved by two students. This means that there are $2 \times 20 = 40$ segments. Let's say there are x students. We know that two segments also leave each blue dot. Therefore there are $2x$ segments. Since $2x = 40$, we have $x = 20$.

(b) Let's show how we can organize the discussion by using the drawing from part (a). Start constructing a string of alternating red and blue dots, where all dots are distinct and neighboring dots are connected by line segments. We can start with some blue dot, then take either one of the two red dots connected with it. This red dot is also connected to another blue dot, and we take this dot which is connected to another red one, and so on. This string cannot be infinite. Therefore, we will eventually reach a dot that is already in the string we are making. The only dot in the string that does not already have two segments attached to it is the blue one we started with, and, because we can't have a dot with three segments attached to it, the string will close up into a loop.

Our string has become a closed cycle in which red dots alternate with blue dots. Separate the cycle into pairs of adjacent dots in either of the two possible ways. There will be a student — a blue dot — and one of the problems he solved — a red dot — in each pair.

If our string does not include all of the dots, do the same with the remaining dots. As a result we will divide all of our dots into several strings,

Problem W.10. (a) A diagonal of a rectangle splits it into two triangles with equal areas. Thus the area of triangle AKD is equal to half the area of rectangle $AKDL$. But the area of triangle AKD also equals half the area of rectangle $ABCD$. This can be seen if we start from point K and draw a line segment KK' that is perpendicular to side AD as in the diagram above. Here triangle $AK'K$ is half of rectangle $ABKK'$, and triangle $DK'K$ is half of rectangle $DCKK'$. Since half the areas of the given rectangles are equal, their areas are equal as well.

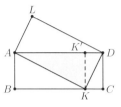

(b) We start by drawing the line segment KD. As in part (a), we can see that the area of triangle AKD is equal to half the area of each of the given rectangles.

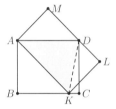

Problem W.11. Because opposite sides of the hexagon are parallel, all six triangles that are adjacent to the hexagon and outside of it are equilateral. The length of any side of the hexagon is one third the perimeter of an adjacent triangle. The sides of all six triangles outside the hexagon form the sides of the original triangles. Therefore the sum of the perimeters of these six triangles is equal to the sum of the perimeters of the original two triangles, which is 21. So the perimeter of the hexagon is equal to $\frac{21}{3} = 7$ cm.

Problem W.12. The answer is yes. First, let us write our six numbers in descending order, $a \geq b \geq c \geq d \geq e \geq f$. Now rewrite them as $acebdf$. Then the difference between the sum of the first three digits and the sum of the last three digits will be equal to $a+c+e-(b+d+f) = (a-f)+(c-b)+(e-d)$. The numbers $c - b$ and $e - d$ are nonpositive, while the number $a - f$ is no larger than 9. Hence the difference $a + c + e - (b + d + f)$ is less than 10.

Problem W.13. We can substitute any integer for x in the expression $ax^2 + bx + c$, and the resulting number will be divisible by 5. Substituting $x = 0$, gives us c. Thus c is divisible by 5. Substituting $x = 1$, gives $a+b+c$. Since c is divisible by 5, $a + b$ is also divisible by 5. Similarly, substituting $x = -1$, we conclude that $a - b$ is divisible by 5. But then $2a$ and $2b$ are both divisible by 5, and therefore a and b are also divisible by 5.

The last sentence requires justification. Why is it that, if a is an integer and $2a$ is divisible by 5, then so is a?

One argument uses the *fundamental theorem of arithmetic*, which says that every positive integer has a factorization into primes, and this factorization is unique apart from the order of factors. Since $2a$ factorizes as $2 \times a$, the prime factorization of $2a$ coincides with the prime factorization of a, apart from an extra factor of 2. So if a is not divisible by 5—that is, if 5 does not occur in the prime factorization of a — then 5 also does not occur in the prime factorization of $2a$.

The theorem just quoted is useful in numerous arithmetic proofs, but it is not easy to prove. So we give another proof that "$2a$ is divisible by 5" implies "a is divisible by 5". Recall that numbers that are divisible by 5 end either in 0 or in 5. If a is not divisible by 5, then its last digit is 1, 2, 3, 4, 6, 7, 8, or 9. But then the last digit of number $2a$ is 2, 4, 6, 8, 2, 4, 6, or 8, and hence it is not divisible by 5. We again reach the desired conclusion.

Problem W.14. In order to find the last four digits of 5^{1000}, let us first calculate a few powers of 5. In fact, we will only calculate the last four digits, and to do so it is sufficient to know just the last four digits of the previous power of 5. We get: $5^2 = 25, 5^3 = 125, 5^4 = 625, 5^5 = 3125, 5^6 = \ldots 5625, 5^7 = \ldots 8125, 5^8 = \ldots 0625, 5^9 = \ldots 3125$. Notice that the last four digits of 5^9 are the same as those in 5^5. Therefore, starting with 5^5, the last four digits of powers of 5 will repeat periodically: 3125, 5625, 8125, 0625, 3125, 5625, 8125, 0625, and so on. Increasing the exponent by 4 does not alter the last four digits of powers of 5. Since the exponent 1000 is divisible by 4, the last four digits of 5^{1000} will be the same as the last four digits of 5^8 which are 0625.

Problem W.15. If an integer is not divisible by 5, then the remainder upon division by 5 is 1, 2, 3, or 4. You should check that, on division by 5, the remainders of $1^4, 2^4, 3^4$ and 4^4 are all 1. Don't forget that instead of multiplying numbers, we can multiply remainders. Thus if a is not divisible by 5, then the remainder of a^4 upon division by 5 is 1. But then $a^{20} = (a^4)^5$ also has the remainder of 1 when divided by 5. Thus the sum of 20th powers of all our numbers will have the same remainder, when divided by 5, as the sum of twenty 1's, which has a remainder of 0.

In this problem we have proved and used the $p = 5$ case of the Little Fermat Theorem which states: If p is a prime and a is an integer not divisible by p, then the remainder when a^{p-1} is divided by p is 1.

Problem W.16. This problem is similar to Problem 11.7, about a bridge across a canal. She must cross both streets, but could we possibly take that into account later? At the moment, the fact that they have nonzero width makes it harder to figure out the best path. If the streets were just

straight lines... Aha! Here is an idea. Let's imagine bringing together blocks (together with the buildings in them), so that one of the streets disappears and becomes a line of width zero, as in the first diagram below. Now do the same with the second street to get the situation on the second diagram. We can draw Carol's path on this diagram and imagine her walking on the diagram from the statement of the problem, but when she reaches a road she immediately teleports to the other side.

It's clear that the shortest path on the modified diagram is simply the line segment HS. Let's mark the points A and B where this line intersects the lines that represent streets, as in the first diagram below, and then pull the blocks apart again, restoring the original road widths and putting the buildings back where they really are. Points A and B will each split into two points — A into points A' and A'', and B into points B' and B'' as in the second diagram below. The dashed path is the shortest path that leads from Carol's home to school.

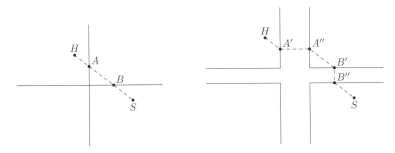

Problem W.17. One answer is the South Pole. Are there any others? It turns out that there are many. Let us build a circular fence of length 1 mile with the center at the North Pole. Consider all points 1 mile south of this fence. If a traveler starts at any such point and walks 1 mile north, he comes to the fence. In walking 1 mile east, he will just walk around the fence and gets back to the same point on the fence. In walking 1 mile south, he returns to the point where he started.

There are even more answers. We can build a circular fence with center at the North Pole and length half a mile or, indeed, of length $1/n$ mile, where n is any natural number. Again consider all points at distance 1 mile south of this fence, and reason as above.

Can you think of any other answers?

Problem W.18. Every time the ray hits a mirror, let's reflect the entire angle across the mirror. We will draw the subsequent path of the ray in this reflected angle as we did in the solution of Problem 12.7. Since the angle of incidence equals the angle of reflection, the path of the ray will turn into a straight line:

Winter Competition

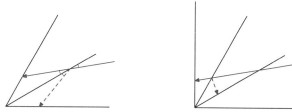

Let us split part of the plane into angles equal to the original angle between the mirrors so that these angles cover the resulting line as in the diagram below. Reflections of the ray will correspond to the intersections of the line with sides of these angles. Clearly, there will be finitely many of these intersections, and therefore, finitely many reflections.

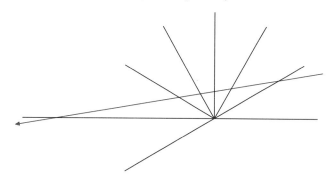

Problem W.19. This problem is very similar to the problem with Winnie the Pooh and Piglet that we have already solved. Suppose the speed of the first tourist, the one leaving from A, is x times the speed of the second tourist. Before meeting at a point C, the tourists spent the same amount of time walking so the distance from A to C is x times the distance from B to C. After the meeting, the second tourist covered a distance x times the distance the first tourist covered after the meeting, and did so at $1/x$ times the speed of the first. Consequently, the time the second tourist spent after the meeting was
$$\frac{x}{1/x},$$
or x^2, times that spent by the first after the meeting. Therefore, $9 = (x^2)(4)$, so $x = \frac{3}{2}$. The distance from A to C is $x = 3/2$ times the distance from B to C so the first tourist spent $x = 3/2$ times as much time walking from A to C as the 4 hours she spent walking from C to B. Consequently, the time she spent walking from A to C is $\left(\frac{3}{2}\right)(4) = 6$. The first tourist walked from sunrise to 12 noon in 6 hours, so sunrise was at 6 am.

Problem W.20. Let's introduce a coordinate system with the unit length equal to the side of a grid square. Then a point will be a lattice point if and only if both of its coordinates are integers. Moreover, every lattice point can be one of the following four types: both of its coordinates are even; both of its coordinates are odd; its first coordinate is even and its second coordinate is odd; or its first coordinate is odd and its second coordinate is even. Since we have marked five lattice points, at least two of them will be of the same type. Thus their first coordinates are of the same parity, and their second coordinates are also of the same parity. This means that the horizontal distance between these two points is even, and the vertical distance between them is even, as well. But then the midpoint of the segment connecting these two points will have integer coordinates, and thus will be a lattice point.

Problem W.21. Suppose at some point in our efforts, exactly n glasses out of the four we plan to flip are upside down. Then after the flip exactly $n - 4$ of the four we turned over will be upside down. The numbers n and $n - 4$ are of the same parity, therefore flipping glasses does not change the parity of the number of upside down glasses, so it will always be odd and thus nonzero. Hence it is not possible to turn all 77 glasses right-side up.

Problem W.22. Let's solve part (b) right away. Since all balls have the same speed, we can simply assume that, after a collision, the balls exchange places as if they were passing through each other with each ball continuing in the same direction as before the collision. Every ball headed in one direction will pass through each of the five balls that are headed in the other direction. Thus there will be a total of $5 \times 5 = 25$ collisions.

Problem W.23. (a) We'll use the same approach as in the solution to Problem 7.6. Let's choose two horizontal lines, one above our triangle and one below the triangle, and start moving these lines toward each other until they touch our triangle. Similarly take two vertical lines, one to the left and one to the right of the triangle, and move them toward each other until they touch the triangle. As a result, we have inscribed the triangle into a rectangle with sides along grid lines. The area of this rectangle is an integer. We can find the area of our triangle by subtracting the area of all extra parts from this integer. All extra parts are either halves of grid rectangles, or consist of halves of grid rectangles and entire grid rectangles as in the examples in the pictures. Thus the area of any extra part can be expressed as half a counting number grid squares. Hence the area of the given triangle can be expressed in the same way.

(b) Suppose, to the contrary, that we have constructed an equilateral triangle with vertices at lattice points. Each side of the triangle connects two lattice points. What is the length of a line segment whose endpoints are lattice points? This line segment is a diagonal of a rectangle with vertices at lattice points so the sides of the rectangle have integral lengths. Using the Pythagorean theorem, we see that the length of the segment is the square root of the sum of the squares of the lengths of sides of our rectangle, and thus it is the square root of a natural number. Let us suppose that the side of our equilateral triangle is $\sqrt{a}$ where a is some natural number. What then is the area of the triangle? Let's draw its altitude. It divides our triangle into two congruent parts which can be rearranged to form a rectangle with side lengths $\sqrt{a}/2$ and h, where h is the altitude of the triangle. Hence the area of the triangle is $(\sqrt{a}h)/2$. We can find h using the Pythagorean theorem: $h^2 + (\sqrt{a}/2)^2 = (\sqrt{a})^2$ so $h = \sqrt{3a}/2$. Therefore the area of the triangle is $(\sqrt{a}/2)h = (\sqrt{3}a)/4$. Since a is a natural number while $\sqrt{3}$ is irrational, the area of the triangle is an irrational number. This means the area can't be an integer or half an integer, which contradicts part (a). Therefore it is impossible to construct an equilateral triangle with vertices at lattice points.

Problem W.24. Assume the opposite, that there exists an arrangement in which it is impossible to cut the board either horizontally or vertically without cutting a domino. If you cut along a vertical board line, it will cut a domino in half. It will also split the board into two rectangles with each rectangle containing an even number of squares because each rectangle has a side of length 6. Take any one of these rectangles. It consists of a number of dominos, and possibly, a number of domino halves. It can't have only one domino half because that would imply an odd number of squares in the rectangle. Thus there are at least two domino halves in each rectangle, so our cut divided at least one additional domino. Therefore, every vertical cut splits at least two dominos. Similarly every horizontal cut splits at least two dominos. Clearly, different cuts, whether in the same direction or not, cannot split the same domino. There are 5 vertical and 5 horizontal cuts for a total of 10 cuts along board lines. Together they split at least 20 different dominos, which is impossible since the total number of dominos is 18. Thus our assumption leads to a contradiction.

Problem W.25. The answer is no. Let's show how Chanelle can play and achieve her goal. She chooses 2^{100} different rows and for 2^{99} turns places an X in each row. After the same number of turns, Erin can place a 0 in no more than half of these rows, so half or 2^{99} of the rows will be free of 0's or "unspoiled". During her next 2^{98} turns, Chanelle will place an X in each of the "unspoiled" rows next to a previously placed X when this is possible. In at least half of these rows, or 2^{98} of them, she will be able to place two X's next to each other, since Erin will be able to "spoil" at most half of the rows. After her next 2^{97} turns Chanelle will have at least 2^{97} rows each containing three adjacent X's, and so on. Eventually she will have at least one row with 100 adjacent X's.

Solutions to Problem Set 14

Problem 14.1. Since a is larger than b, half of a and half of b together is larger than two halves of b and smaller than two halves of a. Thus
$$a > \frac{a+b}{2} > b.$$

Problem 14.2. (a) Gabriela should take the 200 g apple first. Then she would finish eating her apple first and would take the last one. Thus she would be guaranteed to have at least 500 g, while Sam would be able to take just the 400 g apple. If Gabriela takes any other apple to begin with, then Sam would take the 200 g apple and would then end up having eaten more than Gabriela.

(b) This time Gabriela should take the 300 g apple first. Then she would be sure to get 700 g of apple while Sam would get at most 650 g. If Gabriela starts with any other apple, Sam would end up eating more apple.

Problem 14.3. Certainly not. There would a leftmost blue point on the line, and this point cannot be the midpoint of any segment with blue endpoints.

Problem 14.4. (a) Here is a solution. If the numbers were not all equal, there would be one, a, that has a larger neighbor, $a+x$. Then the last number is $a-x$, because only $a-x$ gives $(a+x)+(a-x)/2 = a$. But then $a-x$ is not equal to the average of its neighbors since $(a+x)+a/2 = a+x/2 \neq a-x$. This contradiction shows that the neighbors of a must both be equal to a.

(b) We might, as in part (a), start by denoting one number by a. Then one of its neighbors would be $a+x$, so that the next after this one would be $a+2x$, and so on. One can carry out this argument for a decagon, but it seems long and tedious. It turns out that there is a simpler way to solve the problem by considering the largest among ten numbers. Two neighbors of this number are no larger than the number. But none of them could be smaller than the number since otherwise their average would also be less than the number. Thus these two neighbors are equal to the largest number, and therefore each one of them is itself the largest one. Following the above argument, their own neighbors have to be equal to them, and so on. As a result, we conclude that all ten numbers are equal to the largest number.

Set 14

Problem 14.5. The cuts on the diagram are depicted by heavy lines to intentionally mask the error. If you draw an accurate picture, you'll notice that some parts don't quite fit in the second figure. There will be a narrow gap, as shown below. An extra square was obtained as a result of this.

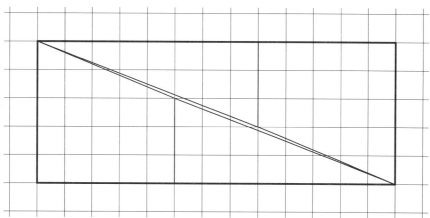

Problem 14.6. (a) Take, for example, weights that weigh 10000, 1000, 100, 10, and 1 kg. Taking some of them and adding their weights, we get a number whose digits are zeros and ones. If the last digit is 1, it means that we have put the 1 kg weight on the balance; if the last digit is 0, the 1 kg weight has not been used. If the second digit from the right is 1, it means that the 10 kg weight has been used, while if this digit is 0 it means that the weight has not been used. This thinking continues on to the fifth place. It is evident that, with different collections of weights, different weights of grain are generated.

One has the option of either placing or not placing each weight onto the pan. Therefore there are $2^5 - 1 = 31$ ways of placing one or more weights on the scales. Each time, the sum of the weights will be different meaning 31 different weights of grain may be recorded.

(b) Your should check that three weights — 1 kg, 2 kg, and 4 kg — are enough. Two weights will be too few because at most three different total weights can be arrived at with two weights. These are one weight, the other, and their sum.

(c) Seven weights — 1 kg, 2 kg, 4 kg, 8 kg, 16 kg, 32 kg, and 64 kg — will suffice. We already proved that any number can be recorded as the sum of different powers of two. In recording a number from 1 to 77 in the form of sums of powers of 2, powers starting with 7 and beyond cannot be used since $2^7 = 128$. Therefore the smaller powers will work, meaning we can use the weights that we have. Actually our weights can measure out 127 different sums: 1, 2, ..., 127. However six weights will not be enough. With six weights, one can measure out not more than $2^6 - 1 = 63$ total weights. It is possible to either place each weight on the scale or not, so the total combinations do not exceed 63.

Think about how the answers in all parts of this problem will change if we are allowed to place weights on both pans, while the grain can only be placed on one pan.

Problem 14.7. (a) Here are three examples:

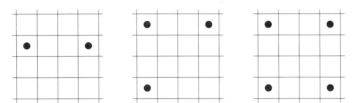

(b) It may seem at first sight that there is no such arrangement. But there is, as one sees on the right.

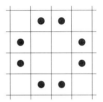

(c) It can't. We will prove it. Let us examine the leftmost vertical column that has a queen, and let us choose the lowest queen in that column. This queen cannot capture to the upper left, or to the left, or to the lower left, or down because there are no other queens in those directions. Only four possible directions remain in which the queen can move. Therefore it can capture at most four queens.

In this problem, we used the important method of examining an extreme case. The most "extreme" queen in this case is the lowest of those that are furthest left. Many problems can be solved by examining an object with the most extreme of a suitably chosen attribute. Examples include the leftmost point in Problem 14.3, or the largest number in Problem 14.4.

Problem 14.8. This problem is similar to Problem 11.7 and Problem W.16 of the Winter Competition. Let A and B be our villages. Let us mark two more points A' and B'. A' is gotten by shifting A toward the first river along a line perpendicular to the river by the length of a bridge across it. B' is gotten from B in a similar way but relative to the second river. The diagram shows these new points. Connect A'

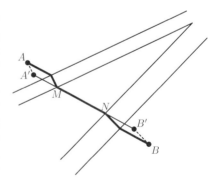

and B' with a line, and let M be the point where this line intersects the bank of the first river which is farther from A. Let N be the point where the line intersects the bank of the second river farther from B. Then one must build the bridge across the first river with one end at point M, and the second bridge must have one end at point N. Try to prove on your own that the length of the path from A to B across these bridges equals the sum of the distance $A'B'$ plus the widths of the two rivers, and that this path is the shortest.

Set 14

Problem 14.9. Yes. Let S be the sum of seven given numbers. If a is any given number, then $S - a$ is a sum of six of the numbers and so must be divisible by 5. If we prove that S is divisible by 5, then we will have to conclude that a is also divisible by 5.

One can choose six numbers from seven given ones in seven ways: taking all numbers but the first; taking all numbers but the second; and so on up to all numbers but the seventh. Add the numbers in each batch of six, and then add the seven resulting sums. Each original number is present in 6 of these batches so, in the end, we will obtain the sum, S, of the original numbers multiplied by 6. On the other hand, we will obtain a number divisible by 5 since each of the seven sums are divisible by 5. But if $6S$ is divisible by 5, then S is also divisible by 5, a fact which you should try to prove.

Problem 14.10. Let's show how we can select ten people that meet our requirement. Let's number the languages of Polyglotia from 1 to 2000, and let's number the population as well. Then let's make a large table of pluses and minuses. The columns will correspond to languages and the rows to citizens. At the intersection of row m with column n, place a plus if the citizen number m knows language number n, and a minus otherwise.

Because more than half the population speaks each language, we know that in each column of the table more than half of the symbols will be pluses. This means that the entire table has over 50% pluses. So then we will be able to find a row where over half of the symbols are pluses as well. Let us pick the corresponding citizen to be the first of our ten. This person speaks more than $\frac{2000}{2}$ languages; therefore, he or she doesn't speak at most 999. Now erase all the columns corresponding to the languages that our chosen citizen speaks. The new table will have at most 999 columns and, in each of the columns, still more than half of the symbols will be pluses.

We can repeat the process. We find a citizen who speaks more than half of the remaining languages; if it happens to be the someone we've chosen before, this will only reduce the number of citizens in our final list. Because we have no more than 999 columns left, this person doesn't speak at most 499 of the 999 languages. Therefore, the first two chosen citizens together fail to speak at most 499 languages.

Repeating the procedure, we choose a third person so that all three collectively fail to speak at most 249 languages. Then we find a fourth, so that the four fail to speak at most 124 languages out of the 2000. Five citizens chosen in this way will fail to speak at most 61 languages; six, at most 30; seven, at most 14; eight, at most 6; nine, at most 2; and ten will collectively have a grasp on all languages.

Problem 14.11. Let's shift each angle to the right, so that its upper vertex coincides with vertex D in the diagram. From the figure we see that the shifted angles together constitute an angle between the diagonal and a side of the square, so the sum we are looking for is equal to 45 degrees.

Solutions to Problem Set 15

Problem 15.1. Of course he can count them out in 60 seconds, but it can be done faster. If he counts out 40 envelopes from one package, there will be 60 envelopes left. This way the salesperson will have spent only 40 seconds.

Problem 15.2. In this solution the juxtaposition of letters indicates decimal notation rather than multiplication; that is, ab is the number $10a + b$.

(a) Let us solve the problem for a five-digit number since the general solution is exactly the same. Let $abcde$ be a five digit number. Let us rewrite it as the sum $abc \times 100 + de$. The term $abc \times 100$ is divisible by 4 because 4 divides 100. Therefore, if de is divisible by 4, then the sum $abc \times 100 + de = abcde$ is also divisible by 4. Additionally, if de is not divisible by 4, then the sum $abc \times 100 + de$ is not divisible by 4 either. But de is the number formed by the last two digits of the given number.

(b) In a similar manner one can prove a criterion for divisibility by 8: An integer is divisible by 8 if and only if the number formed by its three last digits is divisible by 8.

Problem 15.3. (a) Let Sandy's first question be "Is your number bigger than 16?" If the answer is "no", the desired number is one of sixteen numbers: $1, 2, 3, \ldots, 16$. If the answer is "yes", the number is one of the other sixteen numbers: $17, 18, \ldots, 32$. With the next question, Sandy can narrow the search down to eight numbers. In the first case she should ask "Is your number bigger than 8?", and in the second, "Is your number bigger than 24?" Continuing in the same manner, with every question Sandy will cut in half the number of possible choices for Jennifer's hidden number. After her fourth question, there will be only two possible numbers left, and after the fifth question Sandy will find Jennifer's number.

(b) No, she can't. Let's assume, to the contrary, that Sandy has a method of always finding Jennifer's number after only four questions. After each of Sandy's questions, let her write down a plus if Jennifer's answer was "yes", and a minus if the answer was "no". Then after getting all four answers, she will have written a sequence of four signs, some pluses and some minuses. The total number of such sequences is $2^4 = 16$. Using each of these sequences, Sandy has to find a unique number. But then Sandy would be able to find only 16 numbers, a contradiction.

Set 15   135

Problem 15.4. The table is constructed based on the fact that every number can be uniquely represented as the sum of different powers of two. Using numbers 1, 2, 4, and 8, we can get any number from 1 up through 15. How can we find which powers of two should be added in order to get a given number from 1 to 15? Look at the table. If the number is contained in the first row, we include 1 in our sum; if the number is contained in the second row we include 2 in the sum; if it is contained in the third row, 4 must be included in the sum; and if the number is in the fourth row, then we have to include 8 in the sum. For example, if the number is in the first, second, and fourth rows, it is $1 + 2 + 8 = 11$. The number 1 is only in the first row, and the number which is in all rows is $1 + 2 + 4 + 8 = 15$. Try to construct a similar table containing five rows that can be used for finding the numbers from 1 through 31.

Problem 15.5. (See also page 205.) The kids will come back simultaneously five minutes after they turn around. The reason is that the river affects all objects in it — rafts and kids alike — in exactly the same way: it moves them downstream at the rate of the current. But this means that the river can be neglected. Imagine an experiment: we're ridiing a car along the river bank, always keeping opposite the raft, and videotape all that happens. While screening our video, we will see that the raft does not move, and the two kids jump from it simultaneously. In five minutes each one gets some distance away from the raft, and we are not assuming these distances are equal. Then they decide to come back. Clearly, each one will cover the same distance in the second 5 minutes as in the first. (In the language of physics, our experiment corresponds to switching to a frame of reference attached to the river.)

For those not totally comfortable with the solution above, here is an algebraic one. Let v be the speed of the river, and let x be Helen's swimming speed. With respect to a bank, she will swim against the current with the speed of $(x - v)$, and with the current her speed will be $(x + v)$. Therefore, after the first 5 minutes the distance between Helen and the raft will be $5(x - v) + 5v = 5x$. When she turns around and swims with the current, in 5 minutes she will cover a distance of $5(x + v) = 5x + 5v$, so she will catch the raft which, in 5 minutes, has moved a distance of $5v$. The same argument will show that Gary, too, will return to the raft exactly 5 minutes after turning around.

The same will happen if two kids start running in opposite direction in the aisle of a plane moving at a constant speed. If after a few seconds they turn simultaneously around and run towards each other, they will meet at the original spot in the plane. Also if we jump up inside a moving airplane, we'll land on exactly the same spot on the floor of the plane. The airplane moves with a great speed relative to the land, and when we're inside the plane we too move at the same speed relative to the land, but our speed with respect to the plane is 0.

Problem 15.6. (a) Since $10 = 9+1$, and $100 = 99+1$, and so on, a number of the form $10\ldots 0$ will have a remainder of 1 upon division by 9.

(b) Because $a0\ldots 0 = a \times 10\ldots 0 = a \times (9\ldots 9 + 1) = a \times 9\ldots 9 + a$. we can see that the remainder is a if $a \neq 9$, and the remainder is 0 if $a = 9$.

(c) We solve the problem for a five-digit number, since a solution for a number with any number of digits is the same. Suppose we have a number $abcde$ (decimal notation). Let's write it in the form

$$abcde = a \times 9999 + a + b \times 999 + b + c \times 99 + c + d \times 9 + d + e$$
$$= (a \times 9999 + b \times 999 + c \times 99 + d \times 9) + (a + b + c + d + e).$$

The sum in the first pair of parentheses is divisible by 9. Therefore, if the sum in the second parentheses is divisible by 9, the original number is also divisible by 9, and, if the sum in the second parentheses is not divisible by 9, the original number is not divisible by 9. But the sum in the second parentheses is precisely the sum of the digits of the original number.

Problem 15.7. Let's release a light ray from point A so it goes inside the triangle at an angle equal to angle B. This is possible because angle A is larger than angle B. Suppose the ray intersects side BC at a point X. Triangle AXB is isosceles since angles XAB and XBA are equal. Therefore $AX = BX$. But $AX + XC > AC$ by the triangle inequality, so $BX + XC > AC$. But $BX + XC = BC$, and hence $BC > AC$. This is what was to be shown.

Problem 15.8. (a) If the distance between towns A and B is the smallest of all the distances between any two towns, then these towns will have the required property.

(b) Assume, to the contrary, that every town will be someone's destination. Since there are nine towns and also nine people, every town will be the destination of exactly one person. Let us consider two towns, A and B, which are the smallest distance apart. A person from A will head to B, and a person from B will head to A. From above, we know that no other person will head to either A or B. Let us consider the remaining towns and find two towns, C and D, which are the smallest distance apart among these seven towns. Again, a person from C heads to D, and a person from D heads to C, and no one else heads to either C or D. Continuing in the same manner, we'll pick two more pairs of towns, and will be left with one town. There is nobody to head to this last one, a contradiction.

Problem 15.9. In Problem 15.6(b) we proved that if a number is divisible by 9, the sum of its digits is also divisible by 9. But 100! is clearly divisible by 9, so the sum of its digits, S, is also divisible by 9. Also, the sum of the digits of S is divisible by 9, and so is the sum of *its* digits.... Eventually we get a one-digit number divisible by 9. It cannot be 0, so it is 9.

Set 15   137

Problem 15.10. Assume, to the contrary, that at some
airfield A at least six airplanes landed. Draw line segments
from A to the airfields that the airplanes landing at A took
off from. There exist two neighboring segments such that
the angle between them is at most 60 degrees, for otherwise
the sum of all such angles would be more than 360 degrees.
Let these segments be AB and AC. Since all the distances
between airfields are different, all the sides and hence all

the angles of triangle BAC are different. Since angle BAC is at most 60
degrees, the sum of angles ABC and ACB is at least 120 degrees. Thus one
of these angles, say, ACB, is larger than 60 degrees, and hence it is larger
than angle BAC. According to Problem 15.7, the side of triangle which is
opposite a larger angle is longer. Thus we conclude that $AB > BC$. But
then the plane which took off from B could not land at A since airfield C is
closer. We have reached a contradiction.

Problem 15.11. The reasoning is easier to follow if, instead of numbers
from 1 to 16, we had asked that Vivian pick a number between 0 and 15.
Every such number can be written in the binary system using four digits: 0
is written as 0000, 1 is written as 0001, 2 is 0010, 3 is 0011, 4 is 0100, and
so on up through 15 which is 1111. Let's ask Vivian to mentally write her
number in this form, and we will ask our questions about its digits. Our
first three questions will be:

(1) Is it true that the first digit on the left is 1?

(2) Is it true that the second digit is 1?

(3) Is it true that the third digit is 1?

Let the digits that correspond to Vivian's answers be x, y and z, respectively. One of these digits might be false, but our next question will deal with this.

(4) Is it true that the first three digits are x, y and z, respectively?

If the answer is "yes", then we really know the first three digits. In this case the remaining three questions will be the same.

(5) Is it true that the fourth digit is 1?

(6) Is it true that the fourth digit is 1?

(7) Is it true that the fourth digit is 1?

Since Vivian can lie only once, at least two of the answers to these three questions will be the same, and those will be true answers so we will know the fourth digit.

If the answer to question (4) is "no", it means that Vivian has definitely lied once while answering the questions from (1) to (4). Thus from now on she will tell only the truth. The remaining questions will now be the following more cunning ones.

(5) Of the two statements "The first digit is x" and "The fourth digit is 1", is exactly one true?

(6) Of the two statements "The second digit is y" and "The fourth digit is 1", is exactly one true?

(7) Of the two statements "The third digit is z" and "The fourth digit is 1", is exactly one true?

If the last digit is 0, then either all the answers will be "yes" in which case Vivian's number is $xyz0$, and Vivian's answer to question (4) was a lie, or exactly one answer will be "no". For example, if the first of the three answers is "no" then Vivian's answer to question (1) was a lie and her number is $wyz0$, where $w = 1 - x$.

If the last digit is 1, then either all answers will be "no", in which case the number is $xyz1$, and Vivian's answer to question (4) was a lie, or exactly two answers to these three questions will be "no". For example, if answers to questions (5) and (6) are "no", then Vivian's answer to question (3) was a lie, and her number is $xyt1$ where $t = 1 - z$. We see that the answers to questions (5), (6), and (7) first tell us what the last digit is, it's 0 if three or two answers are positive, and it is 1 if one or zero answers are positive.

The last three questions can be stated differently:

(5) Is it true that the sum of the first and fourth digit is x?

(6) Is it true that the sum of the second and fourth digit is y?

(7) Is it true that the sum of the third and fourth digit is z?

By "sum" here we mean the sum modulo 2; namely the remainder on division by 2. So $0 + 0 = 0, 1 + 0 = 1, 0 + 1 = 1, 1 + 1 = 0$.

Using the sum modulo 2, one can devise a cipher which is, in principle, unbreakable. Usually ciphers described in mystery stories are based on replacing each letter by some other letter or symbol. This type of ciphers is very unreliable and here is why. Letters in any English text, and any other language as well, are distributed unevenly. For example, "e" occurs in English texts most often, and "z" the least often. Every letter has its own relative frequency of occurring in an English text, and there are tables of these frequencies. Besides, some letter combinations occur very often, some very rarely, and some, like "qq", never occur. The type of cipher generated by replacing each letter with a corresponding symbol does not change these frequencies, and this leads to a possibility of deciphering. If we find the relative frequency of every symbol in a ciphered text and compare this data with the tables of frequencies, we can guess which symbol corresponds to what letter. Then we can try to replace symbols with letters, verify our hypotheses and make new guesses, and gradually obtain pieces of deciphered text. If the text is long, the probability of getting it deciphered is very high. By the way, the space occurs in a text much more frequently than any letter, so if the space has been replaced by a special symbol, we will have it deciphered first.

Set 15

Let now describe a cipher which, in principle, cannot be deciphered without knowing a special key. Let us pair up each letter of the alphabet with a 5-digit string of 1's and 0's, its binary code. For example, A is 00000, B is 00001, C is 00010, and so forth. The number of these sequences will be sufficient since there are 32 of them, more than the letters of the alphabet. By replacing every letter in a text by its binary code we get a sequence of 0's and 1's, a binary text. This is not yet a ciphered text since it would be easy to decipher it by using a frequency count as described above.

In order to encipher this binary code we will use a key which is simply an arbitrary sequence of 0's and 1's of the same length as the binary text. Both the sender and the receiver of this message will need the key. For enciphering, we add the two sequences — the binary text and the key — digit by digit modulo 2 using the rules $0 + 0 = 0$, $0 + 1 = 1$, $1 + 0 = 1$, $1 + 1 = 0$. The resulting sequence is our cipher. To decipher it, we need to add the key again. Then we will have added the key twice, but according to our rules the sum of two equal digits is always 0, and therefore we get the binary text back. Schematically, the process of enciphering and deciphering can be described as follows: text + key = cipher; therefore cipher + key = text + key + key = text.

It is impossible to decipher a message without the key. If we were given the sum of two numbers, it is impossible to find one of the numbers knowing nothing about the other one. If all we have is the cipher, then all we can tell is that the original binary text could be absolutely any text of the same length. This is so since for any text there is a key which would result in the same cipher!

One disadvantage of this method of enciphering is that every text requires a new key of the same length since repetitive use of the same key increases the chances of deciphering. For example, instead of using a key of the same length, we could try and use a key consisting of just 5 binary symbols, 11010 for example. We break the original binary text into 5-digit symbols, and then add 11010 to each of them. In fact we just replace every 5-digit sequence with another one of the same length. In this case deciphering the text is easy because the cipher is just replacing one 5-digit binary string with another without adding any key. To use longer keys which are still shorter than the text is also dangerous. There are methods of finding the length of the key, and after that has been found, one can use relative frequency analysis. Therefore, a tremendously long key should be prepared ahead of time, and in the beginning of a ciphered text it should be indicated which part of the key is used. It is also important to have a key consisting of a random sequence of 0's and 1's. For example, the sequences 1111111111111111 and 01010101010101010 are not random. By the way, it is not easy to come up with a random sequence. It is possible, but not easy, to clearly define which sequences are considered to be random.

Solutions to Problem Set 16

Problem 16.1. A number is divisible by 72 if it is divisible by 8 and by 9. By the criterion for divisibility by 8, the number 17∗ must be divisible by 8. You can easily check that the only digit that works is 6. According to the criterion for divisibility by 9, the sum of the digits of number 32 ∗ 357176 must be divisible by 9, so we find the last missing digit is 2. Thus the answer is 322357176.

Problem 16.2. (See also page 211.) Draw the escalators as semicircles, and indicate on each semicircle the direction of motion of the escalator it depicts as shown in the diagram. When one of the boys steps off his escalator, he immediately steps on the other one. Thus we can assume that they are running on a moving circle. At first the boys 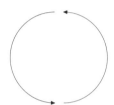 are at one point of this circle while the hat is at the diametrically opposite point. After that the boys start running towards the hat in opposite directions. Since their speeds relative to the escalator are equal, they will reach the hat simultaneously, unless the hat gets to the end of the escalator first. If this happens, the hat will leave the escalator altogether, and thus Jack, the boy who started running down, will get to it first.

Problem 16.3. Not necessarily. The smallest number whose digits sum to 27 is 999, but 999 is divisible by 27 so it is not a counterexample. Can we change 999 so the digits sum to 27 but the number is no longer divisible by 27? We can put 8 and 1 in place of the unit's 9 to get 9981. The sum of the digits of 9981 is divisible by 27 but the number is not, as you can check.

Problem 16.4. This problem will be more easily solved by those who remember what happens when you climb over a fence between two yards. If you started in one yard, then you end up in another one. Let's draw a straight line segment from a point on the baron's land to a point in the village shown on the diagram. Let's mark points of intersection of the segment with the fence. The part of the segment from a point in the baron's land to the first intersection lies in the baron's land; the next part of the segment until the next intersection lies outside his

land; the next part is again in his land, and so on. Therefore if the number of points of intersection is even, both ends of the segment lie in the baron's land, while if the number is odd, then one of the endpoints of the segment does not belong to the baron. There are four points of intersection on our picture, hence the village is on the baron's land.

Problem 16.5. (a) The problem can be solved more easily if we paint the board's squares as on a chess board. Then we notice that bugs which had been on black squares crawl to white squares while bugs which started from white squares will end up on black squares. But our board contains $9 \times 9 = 81$ squares, so there must be more squares of one color, for example more black squares. This means a certain black square will end up empty since there are fewer bugs who start on the white squares to begin with, and they won't be able to occupy all the black squares.

(b) The bugs can change places pairwise. Only the one bug in the corner is an exception and its square will be the only empty one after it crawls off it.

Problem 16.6. A little additional construction will help. We extend the segment AO until it intersects the side BC at a point D. Then $OC < DC + DO$ by the triangle inequality. Also $AO + OD < AB + BD$, again by the triangle inequality. Adding these two inequalities, we obtain $AO + OD + OC < AB + BD + DC + DO$. Canceling DO on both sides and replacing $BD + DC$ with BC, we get the inequality $AO + OC < AB + BC$, as desired.

Problem 16.7. This is similar to Problem 15.8. Let $A, B, C, D, E, \ldots$ be towns visited by Jose in order. By the premises, all distances are different, and Jose always heads to the farthest town from the place he is at the moment. Hence if he came to B from A and then goes to C, distance BC must be larger than AB. If Jose heads to D from C, then $CD > BC$, and so forth. We obtain the chain of inequalities $AB < BC < CD < \cdots$. This chain cannot be infinitely long since there are only finitely many towns, and hence finitely many distances between towns. Thus at some point Jose must come to a town which he has already visited. If this were A and Jose came to A from K, $AB < \cdots . < KA$. But then town K is father from A than B which is a contradiciton since, in the beginning, Jose left from A to B, the town farthest from A. Therefore, Jose will not come back to A.

But how will the chain end? At some point, Jose will leave some town, say M, to go to N, and then he will come back to M. After this he will go back and forth between M and N, having reached either one of the two towns that are the greatest distance apart among all pairs of towns. This is why we had an additional condition that A and C are different towns. Otherwise A and B could have been the two towns which are the farthest from one another, and Jose would go from B right back to A.

Problem 16.8. Rays emanating from a point and meeting one circle form an angle whose measure is less than 180 degrees, as shown in the diagram. Since the sum of two angles less than 180 degrees is less than 360 degrees, two circles are not able to hide a point. Will three circles suffice?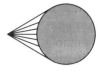

It turns out they will. Starting from the point, let's draw three rays at equal angles to one another as in the diagram below. These rays split the plane into three parts. Let's draw the first circle in such a way that it does not touch the point but intersects two of the rays. As a result, one part of the plane is now hidden. Draw a second circle farther away so that it does not touch either the point or the first circle, but intersects another pair of rays. Now the second part of the plane becomes hidden. Draw a third circle in a similar manner. Now every ray emanating from the given point will meet one of the circles.

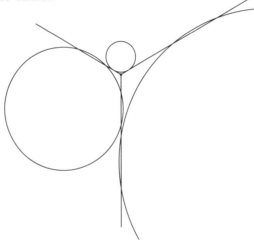

Problem 16.9. Yes, it is possible. Suppose ABC is an equilateral triangle with side 1 and O is a point inside the triangle. Let's connect O with every vertex of the triangle. The triangle will be split into three smaller triangles. The area of triangle AOB is equal to half the product of side AB and the altitude dropped from O to AB. Now $AB = 1$ and the altitude equals the distance from O to AB. Using a similar argument for triangles AOC and BOC, we get the result that the sum of the areas of these three triangles equals half the sum of the distances from O to the sides of the original triangle. On the other hand, the sum of the areas of these three smaller triangles is equal to the area of the original triangle. Thus the sum of the distances from O to the sides of triangle ABC is twice the area of ABC, and therefore it does not depend on the actual position of point O. We can calculate this number using the Pythagorean Theorem. The altitude of an equilateral triangle with side 1 is

$$\sqrt{1^2 - \left(\frac{1}{2}\right)^2} = \frac{\sqrt{3}}{2},$$

Set 16

so its area is $\frac{\sqrt{3}}{4}$, and James must get $\frac{\sqrt{3}}{2}$ as the sum of the distances from O to the sides.

Problem 16.10. This problem was composed by Agnis Andjans, a mathematician from Riga, Latvia. The most remarkable thing here is that a bus stop should be put at point C on the diagram regardless of the lengths of the roads or highway.

First notice that the roads will always contribute the same length to the total distance, so the only important variable is how many times various parts of highway get added to the total.

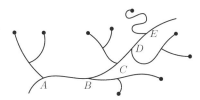

Placing the bus stop to the left of A does not make sense because shifting the stop to A will clearly decrease the total distance. Suppose the stop is located on the highway at some point O in AB. Then two villages are to the left of the stop, and seven villages are to the right. This means that the length AO will be counted in the total twice while the length OB will be counted seven times. Shifting O by 1 mile towards B will decrease the total distance by 5 miles. Before the shift, this 1 mile part of the highway was counted in the total seven times, and after the shift it was counted only two times. Thus shifting the stop all the way to B will certainly decrease the total if the stop were in AB. Now we are on the highway in part BC. Let's start shifting the stop towards C. Each such shift decreases the total distance. For example, shifting the stop 1 mile in the direction of C adds 4 miles to the total distance due to four villages to the left of the stop, but subtracts 5 miles due to five villages to the right of it. Shifting the stop to the right in BC decreases the distance until the stop gets to point C.

Arguing in the same manner, we see that shifting the stop to the right from point C increases the total distance. Therefore the bus stop should be built at point C.

Solutions to Problem Set 17

Problem 17.1.

Problem 17.2. We assume that the price of a packet of tobacco and a box of matches are each a whole number of cents. In this case, the total number of cents in Joe's bill should be divisible by 3. Since 1180 is not divisible by 3, Joe knew that the stated price was wrong.

Problem 17.3. To solve this problem, the chessboard coloring will help us again. Each move of the camel does not change the color of the square it's on. Therefore, it will not be able to get to a square adjacent to its original one.

Problem 17.4. The center of the largest circle is the point where the swimmer was at the moment when that wave was produced. Currently that wave is represented by the largest circle on the diagrams. We can see that the swimmer on the second diagram swims with the speed of the wave. On the first diagram, in the time the wave got from the center to its current position, the swimmer covered only half of the distance. Hence his speed is half the speed of the wave.

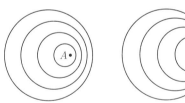

On both diagrams swimmers move to the right. To find the direction the swimmer is headed, one needs to connect the centers of the circles with the point depicting the swimmer.

Problem 17.5. Recall that a line segment connecting the midpoint of a side of triangle with the opposite vertex of the triangle is called its median. It is a theorem that all three medians of a triangle intersect at a point. Let's connect each marked midpoint with the opposite vertex. Now draw a line through the point of intersection of these two segments and the third vertex of the triangle. This line will intersect the third side of the triangle at its midpoint.

Set 17

Problem 17.6. Let's paint the board in a chessboard pattern. In going from room to room the color changes. If it were possible to walk through all of the rooms and return to the starting room without walking through any room twice or leaving the castle, we would obtain a closed chain of alternating black and white squares. Any such chain contains the same number of squares of each color. But the total number of squares in our rectangle is 63 so there are more squares of one color than the other. Thus it is impossible to walk through all of the rooms in the castle as required.

Problem 17.7. One can guess an answer. If the fish surfaced at the point 15 m from the base of the short palm it would surface 10 m from the base of the tall palm. Then the triangle formed by the base and the crown of the short palm and this point is congruent to the triangle formed by this point, the base, and the crown of the tall palm. Thus the birds will get to the fish simultaneously.

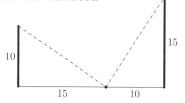

Are there any other solutions? The answer is no. If the fish gets closer to the base of the short palm, then, by the Pythagorean Theorem, its distance to the crown of the short palm will get smaller while its distance to the crown of the tall palm will increase, and so the two distances won't be equal. Similarly, the distances won't be equal if the fish surfaces closer to the base of the tall palm.

Problem 17.8. In this problem, the usual chessboard coloring won't help. Let us color the board in alternating black and white strips, as on the diagram. We will obtain 45 black squares and 36 white squares, 9 fewer than black ones. Because the bugs move diagonally one square, bugs from white squares will end up in black squares and vice versa. Thus at least 9 squares will be empty.

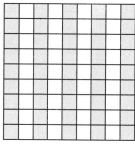

We still have to construct an example where exactly 9 squares become empty. By this is easy: each bug jumps, say, to the next strip to the right, except for bugs on the rightmost strip, which jump to the next strip to the left.

Problem 17.9. Let's write our product as

$$(1!)(2!)(3!)\ldots(99!)(100!) = (2)(2\cdot 3)(2\cdot 3\cdot 4)\ldots(2\cdot 3\ldots 98\cdot 99)(2\cdot 3\ldots 99\cdot 100)$$
$$= (2^{99})(3^{98})(4^{97})\ldots(98^3)(99^2)(100^1).$$

Every number raised to an even power is a square. If we factor out one factor from each number raised to an odd power we'll get

$$(1!)(2!)(3!)\ldots(99!)(100!) = (2\cdot 4\cdot 6\ldots 98\cdot 100)(2^{98})(3^{98})(4^{96})\ldots(98^2)(99^2)$$
$$= (2\cdot 4\cdot 6\ldots 98\cdot 100)\cdot K,$$

with K a perfect square because it is the product of even powers of integers. Now $(2 \cdot 4 \cdot 6 \ldots 98 \cdot 100) = (2 \cdot 1)(2 \cdot 2)(2 \cdot 3) \ldots (2 \cdot 49)(2 \cdot 50) = (2^{50})(50!)$ and 2^{50} is a perfect square. Thus if we cross out $50!$ from the original product we get $2^{50} \cdot K$, which is a perfect square.

Problem 17.10. Part (b) follows easily from part (a), hence it's sufficient to prove (a).

Let's denote the sides of the top left rectangle by a and b, as in the diagram. Suppose that the horizontal side of the top right rectangle is less than a. Then its vertical side is larger than b since their areas are equal. But then the vertical side of the bottom right rectangle is less than a, since the side of the big square would be less than $a+b$ given our assumption. Thus the horizontal side of 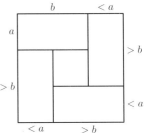 the bottom right rectangle is bigger than b, and so on as we can see in the diagram. Having gone around all the rectangles, we'll have to conclude that the vertical side of the top left rectangle is less than a, a contradiction. Similar reasoning leads to a contradiction if we assume that the horizontal side of the top right rectangle is bigger than a. Therefore, it is equal to a, and so the top right and top left rectangles are congruent. In the same manner we can show that all the rectangles are congruent.

Solutions to Problem Set 18

Problem 18.1. The gears can't turn. Suppose, for example, that the first gear spins clockwise. Then the second spins counterclockwise, the third clockwise, the forth counterclockwise, and so on. Thus the eleventh gear would spin clockwise, and it means that the first one spins counterlockwise which is false since we assumed otherwise.

Amazingly, in three-dimensional space it is possible to make a closed chain with an odd number of gears spinning simultaneously.

Problem 18.2. Let a and b be sides of our rectangle before the change. After the change its sides are equal to $1.1a$ and $0.9b$. The area of the new rectangle is $0.99ab$ which is less than ab, the area of the initial rectangle. The perimeter of the new rectangle is equal to $2(1.1a+0.9b) = 2(a+b)+0.2(a-b)$. It will be bigger than the perimeter of the initial rectangle if $a > b$, and smaller if $a < b$, or equal to the initial perimeter if $a = b$.

Problem 18.3. (See also page 207.)
Solution 1. When the ship is sailing down the river, the current speeds it up compared with the ship sailing on a lake. During each unit of time, the current moves it y additional meters ahead. When the ship is sailing up the river, the current works against it compared with sailing on a lake. During each unit of time it moves the ship y meters back. But the ship sails up and down the river the same distance. Since, in going down it moves faster than going up the river, the current helps it during a shorter period of time than when it works against it. Because of this, going by the river would take longer than going on the lake.

Solution 2. Let the ship's speed be x km/h, and let y km/h be the current's speed. Of course we must assume that $x > y$, for otherwise the ship wouldn't be able to move against the current. With the current, the ship will cover 10 km in $10/(x+y)$ hours, and against it will cover 10 km in $10/(x-y)$. This means that the ship will take the total of
$$\frac{10}{x+y} + \frac{10}{x-y} = \frac{20x}{x^2-y^2}$$
hours sailing on the river. On the lake, the ship will cover 20 km in $20/x$ hours. But
$$\frac{20}{x} = \frac{20x}{x^2} < \frac{20x}{x^2-y^2}$$
because $x^2 > x^2 - y^2$. Therefore the lake round trip would be faster.

Problem 18.4. (a) Each operation reduces the number of numbers written on the board by one. Thus there will be only one number left after nine operations.

(b) Yes, it is possible. Let's notice that the sum of all the numbers on the board becomes one less after each operation. After nine operations, the sum is nine less and, at the same time, it is equal to the number left on the board. At the beginning the sum was equal to $1 + 2 + \cdots + 10 = 11 \cdot 5 = 55$. This means that the number left on the board will be $55 - 9 = 46$.

Problem 18.5. If we count each time water is transferred, each vessel will contain 1 liter of water after every even count. You should notice this if you calculate the amount of water in the vessels from the first transfer.

To prove this in general, suppose that each vessel contains 1 liter and $\frac{1}{n}$th of the water in the first vessel is put in the second vessel. Now the second vessel will contain $1 + \frac{1}{n} = \frac{n+1}{n}$ liters. After that, we need to take $\frac{1}{n+1}$th of the water out of the second vessel to put in the first. The second vessel will then contain

$$\frac{n+1}{n} - \frac{1}{n+1}\frac{n+1}{n} = \frac{n+1}{n} - \frac{1}{n} = 1 \text{ liter.}$$

Problem 18.6. In this game the first player always wins no matter how the second player plays, and, what is most surprising, regardless of how the first player plays. The point is that after each break, the number of chocolate pieces increases by one. At the beginning, there is just one piece; then there are two; then three, and so on until the chocolate bar has been divided into 48 pieces. After the first player's turn the number of pieces is always even, and after the second player's turn the number of pieces is odd. Therefore, the first player, Paul, will win.

Problem 18.7. (a) To draw a route of the mouse as it eats the cheese, it is convenient to divide the big cube into three horizontal layers and draw the mouse's route in each layer separately. The diagrams 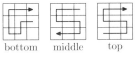 on the right show the three layers for a path that allows the mouse to finish the entire cube except for a corner unit cube.

(b) Let's paint the unit cubes in a checkerboard pattern. The picture shows a coloring of each layer. Now it is easier to count the total number of unit cubes of each color. We can see that with our color scheme there are 14 black and 13 white cubes with the central cube being white. Therefore, the mouse must eat 14 black and 12 white cubes. But if the mouse moves from a cube to a cube only through their shared face, then black and white cubes will alternate along its path. Hence the number of the cubes of different colors eaten by the mouse should either be the same (in the case when the first and the last cubes are of different colors), or differ by one (if the first and the last cube are of the same color). As we have seen above, the number

Set 18

of cubes of different colors that the mouse must eat differ by two. Therefore the mouse will not be able to eat all but the central cube.

Problem 18.8. It is a theorem that a midpoint segment, the segment connecting the midpoints of two sides of a triangle, is parallel to the third side of the triangle. Let's connect the three points that we have. We get a triangle whose sides are parallel to the sides of the original triangle. To reconstruct the original triangle we must draw, through every vertex of the triangle we just obtained, a line parallel to its opposite side. The points of intersection of these lines will be the vertices of the original triangle.

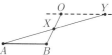

Since this is to be a compass and straightedge construction, we should explain how to draw a line parallel to a given line segment and through a given point using only those tools. It can be done in many different ways, one of which follows. Let O be the given point and AB the given line segment. Find the midpoint of the segment OB and call it X — you should recall how to find the midpoint of a segment with a compass and a straightedge. Draw a line AX and a circle with center X and radius AX. The circle intersects line AX at two points, one of which is A. Call the second point Y. Consider the quadrilateral $AOYB$. Its diagonals, AY and OB, bisect each other at their point of intersection X. Hence $AOYB$ is a parallelogram, and so the line OY is a line through point O parallel to AB as desired.

Problem 18.9. Let's call a 15-minute movement of the snail in one direction a "move". Then we can split all the moves into four types according to their directions: up, down, left, and right. If the snail comes back to the original point, then its path must contain the same number of up and down moves, and also the same number of left and right moves. On the other hand, its vertical (up and down) moves and its horizontal (left and right) moves alternate. Since its path is closed, it follows that the number of vertical moves is the same as the number of horizontal moves. But then the number of left moves is equal to the number of right moves, and to the number of up moves, and to the number of down moves. Hence the total number of moves is a multiple of 4. This means that the snail comes to the original point only after a whole number of hours.

Problem 18.10. (See the diagram.) Let A be the point where the bus is located, and let ray l be the road on which it moves. Let's position our picture so that the road is horizontal and the bus moves to the right. Draw a vertical line, m, through point A. This line splits the plane into three parts: all points to the left of m, all points to the right of m, and all points on m. It is possible to catch the bus only by starting from a point which is to the right of m or at the point A itself. Let's prove this assertion.

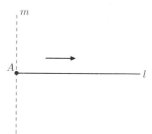

Suppose that we start at a point B and can catch the bus. This will happen at some point of road, call it C. It is clear that we can simply run straight from B to C at the bus's speed and get to C no later than the bus. The set of points from which it's possible to catch the bus at point C is a disc with center C and radius AC. If we draw

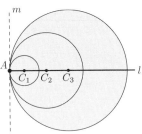

a similar disc for every point of the road, we'll get all points from which it's possible to catch the bus. (The diagram on above shows some of these discs.) None of these discs contains any point which lies to the left of or on line m except point A. This is because the distance from such a point to the center of one of these discs is larger than the radius of the disk. Let's prove this last statement.

Suppose D is a point different from A and on or to the left of m. Let C be on ray l, and consider the disc centered at C with radius AC. In triangle DAC, angle DAC is at least 90 degrees, and thus it's the largest angle of the triangle. Thus $DC > AC$. Hence it's impossible to catch the bus from any such point.

Now suppose that point B lies to the right of line m. Let D be the midpoint of segment AB, and draw the perpendicular bisector of AB. Since the angle between segment DA and ray l is less than 90 degrees, the perpendicular bisector of AB will intersect ray l at some point C. The right triangles CDA and CDB are congruent by SAS, and so segments AC and BC are equal. Thus if we

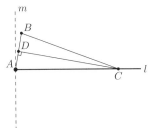

run from B to C at the speed of the bus, we'll get to point C at the same time as the bus.

Problem 18.11. (a) The nickel will make two turns. Suppose the upper nickel on the diagram does not move and the lower one rolls around it clockwise. Notice that the length of the left semicircle of the rolling nickel is equal to the length of the left semicircle of the stationary nickel. Therefore, having rolled over the left half of the stationary nickel, the point of the rolling nickel that started as the bottom will touch the topmost point

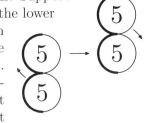

of the stationary nickel, so it will be again the lowest point of the rolling nickel as depicted in the diagram. Thus this nickel assumes the same position as its original one, and hence it has made a full turn. It will complete the second full turn in rolling over the second half of the stationary nickel.

(b) Analyzing the solution of part (a), we see that if the length of the arc along which the rolling nickel touched the stationary one while rolling around it is equal to α degrees, then the angle through which the rolling nickel has turned equals 2α degrees.

Set 18

For example, if the rolling nickel touches the stationary one through 90 degrees or a quarter of a circle, the rolling nickel turns through 180 degrees as shown on the diagram.

What happens when the rolling nickel comes in a position where it touches two stationary nickels at once? At first it was rolling along the first stationary nickel, having gone through an arc of some length, and it is now starting to roll along the second nickel. We can see that on each stationary nickel there is an arc which the rolling nickel does not touch. The length of each of these arcs is one sixth of the circle, or 60 degrees each, as shown in the diagram. These arcs contribute nothing to the rotation of the rolling nickel—only those arcs where the rolling nickel touches a stationary one contribute to the rotation of the moving nickel. As the rolling nickel moves along, there will be one more moment when the rolling nickel touches two stationary nickels at the same time, and therefore two more 60-degree arcs of stationary nickels won't be touched by the rolling nickel.

Let's find the total measure of those arcs of the stationary nickels that were in contact with the rolling one. The total measure of the stationary nickels is 720 degrees. The rolling nickel skips four 60-degree arcs, so that the total measure of all parts of the stationary nickels contacted by the rolling one is $720° - 4 \cdot 60° = 480°$. Hence the rolling nickel has turned through the angle of $2 \cdot 480° = 960°$ with respect to its center. Since $960° = 2 \cdot 360° + 240°$, we conclude that the rolling nickel made two complete turns plus two-thirds of a complete turn.

(c) As in part (b), we must calculate the total measure of the arcs of the stationary nickels touched by the rolling nickel. The total measure of three stationary nickels is $3 \cdot 360° = 1080°$. The rolling nickel skips nine 60-degree arcs, so the total measure of the arcs touched by the rolling nickel is $1080° - 9 \cdot 60° = 540°$. We conclude that the nickel turned through $2 \cdot 540° = 3 \cdot 360°$ around its center, so it made three complete turns.

We can pose a more general problem. Let's place n nickels so that they form a closed chain in which each nickel touches the preceding and the following one. Take one more nickel, and let it roll along the outer edge of the chain without slipping. How many turns around its center will it make if it returns to its starting point? It turns out that the answer does not depend on the chain's shape, the only important thing is the number of nickels forming the chain. Try to prove this statement and find the number of turns made by the rolling nickel.

Problem 18.12. We can pour all the paint from the third jar with equal amounts going into the first and the second ones. Each of these two jars is now full with half as much third-jar paint as its original paint, and, by hypothesis, the paint in each jar is completely mixed. Pour half of the paint in each of the first and second jars into the third jar. As a result, the third jar is now full and contains a mixture of equal amounts of all three paints since it got half of the original amount of paint from the first jar, half of the original amount from the second jar, and two measures of a quarter each, or half, of the original amount of the third paint. Now pour the entire contents of the second jar into the first jar. It now also contains equal amounts of all three paints, since it has the rest of all the paints. Now we just pour one third from each of the first and the third jars into the second jar, and we're done.

Solutions to Problem Set 19

Problem 19.1. It suffices to open three links of one of the pieces. Then the remaining four pieces can be connected using these three links.

Let's prove that it is impossible to reattach the chain if only two links are opened. Indeed, if both of these open links come from the same piece, we have to use them to connect four other pieces and one unopened link, which is impossible. If the two open links came from two different pieces, then we have to use them to connect five pieces of chain together, which is impossible.

Problem 19.2. (a) Yes: for example, $1 + 2 - 3 + 4 + 5 + 6 - 7 - 8 = 0$.

(b) No, it is not possible. We give two proofs.

Solution 1: Suppose we have found a suitable replacement for the asterisks and obtained an equality. Let us move all numbers with a minus sign to the right-hand side of the equation. As a result, the sum of some of the numbers $1, 2, 3, \ldots, 9$ which are on the left side is equal to the sum of the rest of them which are on the right. But this means that the total sum of these numbers is even. This is false since $1 + 2 + \cdots + 8 + 9 = 45$.

Solution 2: The problem can be stated in the following form: Is it possible to replace the asterisks in the expression $1*2*3*4*5*6*7*8*9$ by plus and minus signs so that the value of the resulting expression is zero? Let's first replace every asterisk by a plus. We get an odd number, $1+2+\cdots+9 = 45$. Now let's replace one of the pluses by a minus. How does the value of the expression change? If we change the sign in front of a number a, then instead of the expression $1 + \cdots + a + \cdots + 9$, we'll get $1 + \cdots - a + \cdots + 9$, whose value is $2a$ less than the value of the preceding one. Thus the new value differs from the old one by an even number, and hence the value of the expression is still odd. Regardless of the number of times we replace a plus with a minus, or vice versa, the value of the expression always changes by an even number, and hence it always remains odd. Therefore, we can never make it equal zero.

Problem 19.3. Let ABC be our triangle, and let M be the center of the circumscribed circle that coincides with the midpoint of side AB. Let's prove that angle C is a right angle. Triangles AMC and BMC are isosceles; hence $\angle A = \angle MCA$ and $\angle B = \angle MCB$. But the sum of the angles MCA and MCB equals angle C. Thus the sum of angles A and B equals angle C, so angle C is half the sum of angles A, B, and C. Since the sum of the angles of a triangle is 180 degrees, angle C is 90 degrees.

Problem 19.4. Suppose we have placed numbers in the table and gotten the desired results. We can calculate the sum of all numbers in the table in two different ways. One way is to find the sum for each column and then add these sums, and another is to find the sum for each row and add these sums. We must get the same number. Since the sum for each column is positive, the total must be positive. On the other hand, the sum for each row is negative, so the total must be negative. Since no number is, at the same time, positive and negative, we have a contradiction. Hence placing numbers in the table with row sums negative and column sums positive is not possible.

Problem 19.5. (a) Here are some examples:

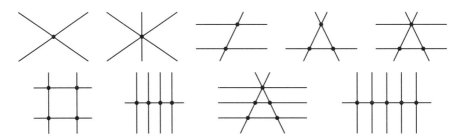

(b) We can determine the number of lines only for the situation with two points of intersection. In this case Jamie's picture must contain exactly three lines. Let's prove it. Take one of these points and call it A. Two lines, l_1 and l_2 intersect at A. Since there is one more point of intersection—call it B—there must be

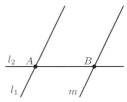

a third line, m, not passing through A. If m is not parallel to either l_1 or l_2, we get at least three intersection points, which is a contradiction. Thus it's parallel to, let us say l_1, and intersects l_2 at B. We have a picture with three lines, and we can't add any more lines without getting additional points of intersection. If a new line passes through A and does not coincide with either l_1 or l_2, then it must intersect m at a new point. If a new line passes through B and does not coincide with either m or l_2, then it must intersect l_1 at a new point.

Other pictures may contain different number of lines, as can be seen from the examples in part (a). Generally, for any number $N > 2$ of points, one can draw examples with different numbers of lines. This is particularly obvious for the arrangement of points on the right. Depending on whether or not we include the dashed line, we have either N or $N + 1$ lines.

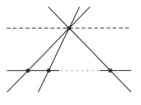

Problem 19.6. (a) A 24-hour day starts at 00:00, at which time we have the first coincidence. The hour hand moves at the rate of $\frac{1}{12}$ of a complete revolution per hour; the minute hand moves at a rate of 1 complete revolution per hour, and the hands move in the same direction. One can think of this as the minute hand chasing the hour hand and catching up at a rate of $1 - \frac{1}{12} = \frac{11}{12}$ complete revolution per hour. To see this, let's think of the hour hand as being stationary — as it's said in physics, we switch to a frame of reference attached to this hand. We can put the clock horizontally on the table in front of us, and continually turn it so that the hour hand stays in the same position. The minute hand will then turn around the hour hand at a rate of $\frac{12}{11}$ revolutions per hour. Think of the minute hand as running laps around a track and completing each lap in $\frac{12}{11}$ hours. In 24 hours it will complete

$$\frac{24}{12/11} = 22$$

laps. It starts at the start/finish line and crosses the start/finish line 21 more times, and at the end of the 22nd lap it's back at the start/finish line. So there are 22 coincidences on May 17. (The one at the end of the 22nd lap, 24:00 — that is, 00:00 on May 18 — does not count.)[1]

(b) As in part (a), let's think of the hour hand as not moving. Then the minute hand will rotate around it at a rate of $\frac{11}{12}$ revolutions per hour. During a 24 hour period, it will go around the clock 22 times as we've seen in part (a). During each complete revolution, the minute hand forms a right angle with the hour hand twice. Thus, during 24-hour period, it happens 44 times.

Problem 19.7. (See also page 208.)

Solution 1: Let's declare the distance between A and B to be 1 — it actually is if the unit of distance is the distance between A and B. Let the speed of the boat be v, and the speed of the current be x. According to the statement of the problem, $3 \cdot (v+x) = 1$ and $5(v-x) = 1$. Thus $v+x = \frac{1}{3}$ and $v-x = \frac{1}{5}$. Subtracting the second equation from the first, we get $2x = \frac{1}{3} - \frac{1}{5} = \frac{2}{15}$, which implies $x = \frac{1}{15}$. Hence a raft will take 15 days to get from A to B.

Solution 2: Let's simultaneously release a raft and two identical motorboats from point A, with one boat going downstream and one going upstream. The raft will always be exactly half way between the two boats. In the time the boat going downstream covers the distance of $5t$, the boat which is going upstream covers the distance of $3t$. Hence, at that moment, the midpoint

---
[1] Once this problem was given to participants in the National Russian Mathematics Correspondence School. One of the students sent in a right answer, but instead of a reasoned solution, he just wrote what he did in order to get an answer — he simply sat for 24 hours looking at his wristwatch and counted the times when hands coincided. The person who graded his test wrote back to the student saying he could have obtained the answer much faster by turning the hands of his watch manually, until the hour hand had made 24 revolutions!

between two boats, which is the raft's position, will be a distance t from A in the direction of B. In order for the raft to be at point B, we should take t to be equal to the distance between A and B. By that time the downstream boat will have covered a distance five times that of the distance between A and B, which means that it took the raft 15 days.

Problem 19.8. Let A be the point where the bus is and l the ray representing the road the bus rides: (We position our picture so that the road is horizontal and the bus moves to the right.) Let's draw rays m and n, which are based at A, make 30-degree angles with l, and lie on different sides of it, as shown on the left. These rays form a 60-degree angle. The bus can be caught only if one starts from a point inside this angle or on one of the rays m or n. It is impossible to catch the bus starting from any other point. Let's prove it.

Suppose we're at a point B in the field, and suppose we can catch the bus at some point on the road, say C. It is clear that, if we run directly from B to C at 5 mph, we will get to C no later than the bus. But the set of points from which it's possible to catch the bus at the point C is the disc with center at C and radius $AC/2$, since the bus's speed is twice that of someone running to catch the bus. So far we have shown that the set of points in the field from which one can catch the bus is the set of points gotten by drawing a disk of radius $AO/2$ about each point O on the road. Now we will prove this is the same as the points on and between the rays m and n.

Let's prove that if O is a point on l, the disk of radius $AO/2$ centered at O lies between the rays m and n. Indeed, take O on l and draw the segment OX from O that is perpendicular to ray m at point X. In the right triangle AXO, the angle AOX equals 60 degrees, and therefore $OX = AO/2$. If we draw a circle of radius $AO/2$ about O, it will go through X and the radius OX will be perpendicular to m. This means that m is tangent to the circle, and so the circle lies on one side of m. Similarly, the disc lies on one side of the ray n, and hence the disk is inside our angle.

Now let's prove that if we start at any point of our angle it's possible to catch the bus. Suppose we're at a point B which lies between the rays m and l. Draw a line through B which is perpendicular to m. This line intersects rays m and l at points X and O, respectively. We will catch the

Set 19   157

bus if we run from B along BO because $AO = 2OX$ and we would get to O simultaneously with the bus even if we start at the point X.

Problem 19.9. For any vertical line that does not pass through a digit, consider the difference between the number of 1's to the left of the line and the number of 0's to the right of the line. If the line is to the left of the number, there are no 1's to the left and at least one zero to the right, so the difference is negative. Let's shift the line to the right digit by digit, and analyze how the difference changes. If we jump over a one, the number of 1's to the left of the line increases by one, while the number of 0's to the right of the line does not change. Thus the difference increases by 1. If we jump over a zero, the number of 1's to the left of the line does not change, while the number of 0's to the right of the line decreases by 1. Hence the difference again increases by 1. Thus, we start with a difference which is a negative integer, and the difference increases by at most 1 at each step. But when we have shifted the line to the very end and placed it to the right of the number, the difference is positive since there is at least one 1 to the left of the line and the number of 0's to the right of it is zero. Therefore, at some moment the difference was 0.

Problem 19.10. Let ABC be our triangle, and suppose the ant sits at the midpoint M of side AB. The ant's path is a closed curve consisting of three line segments. The first segment connects point M and some point X of side BC, the second segment connects X and some point Y on side AC, and the 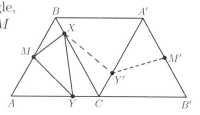 third segment connects points Y and M (see figure). Let's reflect our triangle together with the second segment XY across side BC. We get a triangle BCA' and a point Y' on side CA'. The path of the ant from M to X to Y consists of segments of the same length as the path from M to X to Y'. Let's now reflect triangle BCA' across side CA'; we get a triangle $CA'B'$. Let M' be the midpoint of side $B'A'$. The ant's path from M to X to Y to M consists of the same length segments as the path from M to X to Y' to M', so the paths have the same length. But the shortest path from M to M' is the line segment MM'. This segment intersects sides BC and CA' at their midpoints. Hence the ant should first move straight to the midpoint of BC, then directly to the midpoint of AC, and then back to M.

Problem 19.11. The first steps in the solution of this problem are virtually automatic simply because we have almost no other choice of action. What can be done when the tools are a straightedge and a pencil? All we can do is connect points in the figure with straight lines, so connect the point M and the endpoints, A and B, of the diameter with straight line segments. Suppose that MA intersects the circle at a point X, and that MB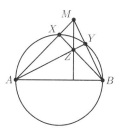

intersects the circle at a point Y. What do we do now? It seems that we don't have much choice again; the only possibility is to draw segments AY and BX. These two segments intersect at a new point, Z. Now draw a line through points M and Z. Now we should stop since the problem is solved. Indeed, angles AXB and AYB are right angles since they are inscribed in a circle, and AB is a diameter. Hence Z is the point of intersection of altitudes BX and AY of triangle AMB. But all three altitudes of a triangle intersect at a point. Thus line MZ is perpendicular to AB.

For the point N we do exactly the same, only the altitudes will intersect outside the triangle ANB, and the perpendicular from point N will fall not on the diameter AB but on its extension.

It turns out that the problem can be solved regardless of the position of the point M. It can lie outside or inside the circle, on its diameter, on the circle itself, or even at one of the points A or B. Solving the problem might not be as easy in some of these cases, but nevertheless, try to analyze all of the possible situations.

Problem 19.12. The answer is no. Here is an unusual coloring of the board that will be helpful. Let's split the 10×10 board into twenty-five 2×2 squares. Color these 25 squares in a checkerboard pattern. Each 4×1 rectangle will then cover exactly two white and two black squares. Thus if we could pave the board with such rectangles, the board would have as many white squares as black ones. But with our coloring, this is 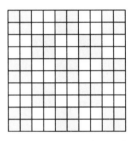 not so, as you should check. Therefore it's impossible to pave the board with 4×1 rectangles.

Problem 19.13. The first player can guarantee himself a victory. To begin with, he moves the marble 19 squares to the left, to the opposite end of the grid. Afterward, he moves the marble to the end of the board in the direction where his opponent just moved it. Let's prove that the first player will always be able to move. Suppose that the second player moved the marble through x squares and the move of this length had not been done prior to that. Then the first player would need to move the marble through a different number of squares. The numbers x and $19 - x$ are different since they have different parity. This is a nonzero shift since $x \neq 19$. A shift of this length has never occurred before either, since otherwise the shift of length $19 - (19 - x) = x$ would have occurred earlier, but this is not the case.

Solutions to Problem Set 20

Problem 20.1. There are 18 seconds between the first and the thirteenth ticks. Drawing a diagram can help because the numbers are not very large. Lightly mark a point at every centimeter with a pencil on a line to represent seconds. Next, starting with the first point, mark every second point with green ink to represent ticks of the first clock. Beginning with the first point again, mark every third point with red ink to represent ticks of the second clock. Count 13 points marked with ink of any color and find the number of centimeters between the first and the last. This is the number of seconds between the first and the thirteenth ticks.

This method always works, even if we need to find the number of seconds between the first and the thousandth tick, but it may take a lot of time. Therefore let's find another solution. Since the clocks started to click at the same time, the first tick we hear is both clocks ticking. The ticks will coincide every 6 seconds, and between coincident clicks there are 3 single clicks. Therefore we can group 4 clicks in 6-second time intervals, but we must remember that we are counting coincident clicks as coming before 3 single clicks, rather than at the end of them. Because 13 is 3 times 4 and 1 more, so 3 times 6, or 18 seconds, will pass in the time that we hear 12 ticks and 1 more, the 13th click, occurs at the end of the 18th second.

How much time will pass between the first and the thousandth ticks? Because $1000 = 250 \cdot 4$, there will be $250 \cdot 6 = 1500$ seconds between the first and the 1001st ticks. If we start and end with consecutive coincident clicks, the time intervals between ticks are 2, 1, 1, and 2 seconds. We do not need to count the 1001st tick, which is the last in the last set of 6. Since the previous tick is 2 seconds before it, there are 1498 seconds between the first and the thousandth ticks.

Problem 20.2. The volume of the coffee in the first cup equals the volume of the milk in the second cup. After transferring one teaspoon of liquid from the first to the second cup and then one teaspoon from the second to the first cup, each cup again holds 150 ml of liquid. If the first cup now contains x ml of coffee, then it contains $150 - x$ ml of milk. Therefore, the remaining x ml of milk and $150 - x$ ml of coffee are in the second cup.

Now solve the problem if the two initial volumes of liquid were different, but the volumes transferred were the same as in the original problem. Does the answer change?

Problem 20.3. Sides AB and DC are not parallel, because otherwise we would have a parallelogram in which opposite sides are equal, contrary to the hypothesis that $AD > BC$. Extend sides AB and DC to their intersection at some point X. Because $AD > BC$, sides AB and DC will intersect beyond points B and C as you can see in the diagram. Thus the sum of angles A and D of the trapezoid is equal to $180 - \angle X$, so it is less than 180 degrees. The sum of angles B and C of the trapezoid is equal to $\angle ABC + \angle BCD = (180 - \angle XBC) + (180 - \angle XCB) = 360 - (\angle XBC + \angle XCB) = 360 - (180 - \angle X) = 180 + \angle X$, so it is greater than 180 degrees. Therefore, the sum of angles B and C is greater than the sum of angles A and D.

Problem 20.4. Let the notepad cost x cents. There are 100 cents in one dollar, so the problem can be rewritten as two inequalities: $9x < 1000$ and $10x > 1100$. From the second inequality we find that $x > 110$. If $x = 111$, the first inequality is true as well because $999 < 1000$. If $x > 111$, so that x is at least 112, then the first inequality is no longer true. Therefore only one solution is possible and the notepad costs 111 cents.

Problem 20.5. Let AX and BY be the two heights of triangle ABC we are dealing with. According to the statement of the problem, $AX \geq BC$ and $BY \geq AC$. Because AX is a height, $AX \leq AC$, and because BY is a height, $BY \leq BC$. The first pair of inequalities implies that $AX + BY \geq BC + AC$, while the second pair of inequalities implies that $AX + BY \leq BC + AC$. Therefore $AX + BY = AC + BC$. Now $AX \geq BC$ and $AX + BY = AC + BC$ imply $BY \leq AC$. However $BY \geq AC$, by hypothesis, so $BY = AC$ and $AX = BC$. Also $AX \leq AC$ and $AX = BC$ implies $BC \leq AC$, and $BY \leq BC$ and $BY = AC$ implies $AC \leq BC$. This shows that $AX = BC = AC = BY$. Therefore our triangle is a right isosceles triangle with $\angle C = 90°$, which leads to $\angle A = \angle B = 45°$.

Problem 20.6. Let's not pay attention to recruits who turned around. Imagine the recruits in front of the sergeant as a row of 0's—corresponding to recruits who turned left—and 1's—those who turned right.

Now the problem of finding a place for the sergeant can be formulated as follows: Where can we draw a line so that the number of 0's to its right equals the number of 1's to its left? The solution will now be very similar to the solution for Problem 19.9. In moving the line from left to right we will find a place for which the number of 0's on the left is equal to the number of 1's on the right.

Problem 20.7. This solution is similar to the second solution for Problem 19.2(b). Notice that $1 + 2 + \cdots + 10 = 55$, which is odd. When we erase two numbers, a and b, and replace them with $a - b$, the overall sum of the numbers on the board does not change from odd to even or vice versa. We simply replace a and b, which contribute the value of $a+b$ to the total sum, with one number $a-b$, thus decreasing the total sum by the even number $2b$. Therefore, no matter how many times we conduct this operation, the sum of the numbers on the board will remain odd. Hence we can never end up with 0, or any other even number.

Problem 20.8. Let x be the number of students in the circle. We need to find the least x possible such that between the two numbers $48x/100$ and $50x/100$ there is an integer. This integer will be the number of girls in the circle. We can solve the problem by trial and error, consecutively trying to check $x=1, x=2$, and so on. But this is long and tedious.

The problem can be solved more easily if we examine two cases: when x is even, and when it is not. If x is even, then 50% of x is an integer. To satisfy the conditions of the problem, there must be an integer between 48% of x and 50% of x. Therefore we need to find the least even x for which $50x/100 - 48x/100 > 1$. This inequality reduces to $2x > 100$, so $x > 50$. The smallest even x that satisfies the last inequality is 52.

If x is odd, then 50% of x is halfway between two integers. Therefore we need to find the least odd x such that $50x/100 - 48x/100 > 1/2$. This inequality reduces to $2x > 100/2$, so $x > 25$. The smallest odd x larger than 25 is 27. Comparing the two numbers, we conclude that the least possible x is 27.

Try to solve the problem if we know that the number of girls in the math circle is less than 44% but greater than 43%.

Problem 20.9. This is very similar to Problem 19.8. At first the man is standing on the road. Let's say he needs to get somewhere in 1 hour or less. If he leaves the road at some point, then he should not return to the road because if he leaves the road at some point and returns to the road at B, he could have saved time by going along the road directly to B. Therefore the quickest way from point A to any other point consists of either moving along the road and then along the field straight to the destination, or straight to the destination when it is on the road. Now we can slightly reformulate the problem: From what points is it possible to reach point A in 1 hour if you first move along the field straight to some point on the road, and then along the road to A?

Now we can continue as in Problem 19.8. Draw circles centered at each point on the road. If the point on the road is located some distance, s, from A, the radius of the circle should be equal to $(6-s)/2$, so that it is possible to reach A in one hour from all of the points in the disc by first walking to the center of the circle and then along the road to point A. The shape we are looking for is the union of all of these discs.

We can also almost reduce the problem to Problem 19.8. The point A divides the road into two parts, the left and right sides. We need only solve the problem when we start to the left of A because a solution will be symmetric on the right side. Let's find a point, M, on the road to the left of A, from which we can reach A by moving along the road in one hour. Send a bus that moves at 6 km/h out of this point towards A. We now need to find all of the points from which we can catch the bus somewhere between M and A. This will be a part of the 60 degree angle with its vertex at point M bisected by the road. To find what part of the angle we are looking for, remember that we need the union of the circles whose centers are on segment MA such that the radius of the circle at point X on the road is equal to $MX/2$. The last of these circles has center A and radius $MA/2$ (since $MA = 6$, and so $MX = (6 - AX)/2 = (6 - s)/2$, where $AX = s$).

Draw tangent lines from M to this circle. This is the part of the angle from which it is possible to catch the bus before it passes A; it is the region bounded by the angle and the last circle.

We can solve the problem for the right side of the road analogously. By combining all the points we found, we end up with the following picture.

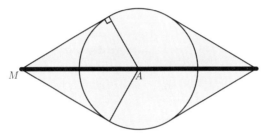

Problem 20.10. This problem is different from other game problems we have seen. It does not admit any strategy we have used for other games. But the problem can still be solved in a manner that is easy to understand if not to discover. The second player can win no matter how the first player plays. After the first player paints a square, the second player should paint her pattern near one of the corners of the board, as shown in the diagram. This is always possible because at least 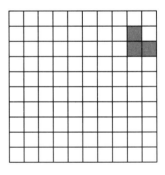 three corners are still available after the first move. We see that between the colored group and the edge of the board there are now three squares forming a second-player pattern—squares that the first player is no longer allowed to paint! So the second player will "save" these squares (avoid painting them) as long as possible. Apart from saving these squares, she can play anywhere. If at some point her only available moves involve painting a saved square, the first player has no moves at all. So the second player paints the saved squares and wins.

Problem 20.11. Let's analyze how the number of pieces can change after a given cut. To begin with, there is 1 piece, and after the first cut there are at most 2 pieces. After the second cut, there can be at most 4 pieces,

Set 20   163

which we will get if we stack the previous 2 pieces before cutting. Likewise the maximum number of pieces after the third cut is 8, which we obtain by stacking the 4 previously obtained pieces before cutting. Continuing in the same manner, we find that the maximum number of pieces after the fourth cut is 16, it is 32 after the fifth cut, and it is 64 after the sixth cut. Since this is the number of pieces we need to obtain, we see that it is impossible to produce 64 unit cubes using fewer than six cuts. On the other hand, it is possible to do using exactly six cuts, and we will leave it to you to find a way to do it. Our reasoning so far provides a hint; the number of pieces must double with each cut.

The fact that we need at least six cuts can be proved using a different approach. Let's consider some inner $1 \times 1 \times 1$ cube. It must be cut out, which means that we need to make cuts along each of its six faces, and all these cuts will be different. We have used this idea in solving Problem 8.8.

Solutions to Problem Set 21

Problem 21.1. Let's find a number whose square is not much different from 40500. Take 200; its square is 40000, which is less than 40500. The square of 201 is 40401, also less than 40500. The square of 202 is 40804, which is greater than 40500. The answer is 40804.

Problem 21.2. (a) Notice that the difference $a^2 + b^2 - 2ab$ is equal to $(a-b)^2$, and is nonnegative because the square of any number is nonnegative. So $a^2 + b^2 \geq 2ab$ for any a and b.

(b) Because $x > 0$, we can multiply both sides of the inequality $x + 1/x \geq 2$ by x, thus getting an equivalent inequality $x^2 + 1 \geq 2x$. We can see this last inequality is true because $x^2 + 1 - 2x = (x-1)^2 \geq 0$.

Alternatively, as you should check, we can immediately reduce the given inequality to the one we proved in part (a) if we set $a = \sqrt{x}$ and $b = 1/\sqrt{x}$.

(c) By multiplying both sides of the given inequality by the positive number b, we reduce it to the inequality we proved in part (a).

Problem 21.3. (a) Yes; for example, $-2, -2, 5, -2, -2, 5, -2, -2, 5, -2, -2$. (Check that it works.)

(b) No. Divide our sequence into four sets of three numbers each. The first set consists of the first three numbers, the second set consists of the next three, and so on. The sum of each set will be positive, and the sum of all 12 numbers is equal to the result of adding the four sums from the individual sets. The sum of all the numbers is a sum of four positive numbers, which must itself be positive.

Problem 21.4. (a). The sum of two adjacent sides of a rectangle is one half of its perimeter so, in this case, is equal to 2. Let one of the sides equal a. Then the side adjacent to it is equal to $2 - a$. The area of the rectangle is then equal to $a(2-a)$, which is at most 1 because $1 - a(2-a) = a^2 - 2a + 1 = (a-1)^2 \geq 0$.

We could also solve the problem a bit differently. Adjacent sides of the rectangle can't be shorter than 1 because their sum would be less than 2. Therefore we can let the length of one of the sides be $1 + x$ with $x \geq 0$. Then the other side has length $1 - x$ because the sum of adjacent sides is equal to 2. The area of the rectangle is then equal to $(1+x)(1-x) = 1 - x^2$ which can't be larger than 1.

(b) This is possible, but only when the rectangle is a square of side 1. This follows from part (a).

Problem 21.5. Let I be the point of intersection of the angle bisectors of triangle ABC. Because MN is parallel to side AC, angles MIA and CAI are equal. Angle CAI is equal to angle MAI because AI is the bisector of angle A. Therefore angles MIA and MAI are equal, so AMI is an isosceles triangle and therefore $AM = MI$. Similarly we can prove that triangle CNI is isosceles and therefore $CN = NI$. Therefore $AM + CN = MI + NI = MN$, and we have completed our proof.

Problem 21.6. Solution 1: Let's find out how many times the cashier counts a given bill. For example let's check how many times a particular $10 bill gets counted. It gets counted first when the cashier counts the total number of bills, it is counted again when he counts the bills greater than $1, and so on. The last time he counts the $10 bill is when he counts the number of bills greater than $9 in value. Thus this bill gets counted exactly 10 times. Similarly he counts each $n bill exactly n times. This means that the cashier will end up with the correct dollar amount.

Solution 2: Let's represent an $n bill as a column of n unit squares. Place the columns representing all the bills next to each other in ascending order of heights. We end up with a figure consisting of unit squares like the one depicted in the diagram. The total number of squares in the figure is the total sum. But how does the cashier count the money? First he counts the total number of bills—this is the number of squares in the lowest row of our figure. Then he counts the number of bills greater than $1 in denomination—this is the second row, and so on. As a result the cashier will count the total number of squares in the figure, and thus will find the total dollar amount.

Problem 21.7. Using the results of Problem 21.2(a), write down the three corresponding inequalities: $x^2 + y^2 \geq 2xy$, $y^2 + z^2 \geq 2yz$, and $z^2 + x^2 \geq 2zx$. Adding these inequalities, we get $2x^2 + 2y^2 + 2z^2 \geq 2xy + 2yz + 2zx$. After dividing both sides by 2, we end up with what we were asked to prove.

Problem 21.8. (a) Yes. Let's see what will happen if you take one coin from the left table and transfer it to the right while flipping it over. If the coin was on the left table tails up, it will end up on the right table heads up. There will now be the same number of heads on the left table as there were originally, and one more on the right. If the coin was heads up on the left, it will land tails up on the right. Now the number of heads on the left will decrease by 1, while the number on the right will not change. Therefore after each such move, the difference between the number of coins lying heads up on the left table and the number heads up on the right will decrease by 1.

Having done this 31 times, you will make the difference zero, and will have achieved what was originally asked.

(b) This problem is solved similarly, except that you should only transfer and flip 23 coins.

Problem 21.9. This problem is solved similarly to Problem 20.9. Here are the answers:

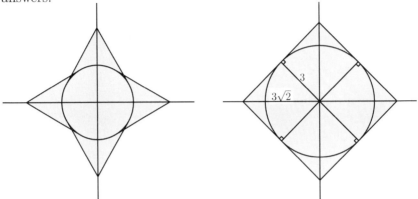

Problem 21.10. To solve this problem, one needs to know some physics. A beam balance can be thought of as a straight rod AB with pans tied to the ends. The rod itself can pivot around its fixed point C. If C is the midpoint of AB, then the balance works correctly provided the

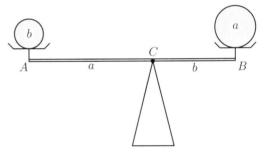

pans are of the same weight, which we assume for this problem. If the point C is misplaced, the balance will be false. Suppose C divides the rod in proportion $a : b$ from left to right. Then the loads whose masses are in proportion $b : a$ from left to right will balance. If we place a 1 kg weight on the left pan, it will be balanced by the load whose mass is a/b kg placed on the right pan. If we place a 1 kg weight on the right pan, we need to place a load of mass b/a on the left pan to balance it. Thus the farmer will have weighed out $(a/b + b/a)$ kg, which is always greater than 2 when $a \neq b$; compare with Problem 21.2(a).

To weigh 2 kg of nuts, first place the 1 kg weight on the left pan and balance it with nuts on the right pan. Then remove the weight and replace it with nuts on the left pan so that they balance the nuts already on the right pan. Now we have exactly 1 kg of nuts on the left pan. Repeating this procedure we get 1 kg more.

Problem 21.11. Notice which square holds the number 1 and which holds the number 64. These squares are either in one row, or in one column, or in opposite corners of a rectangle. In any case it is possible to move from

Set 21

the square with number 1 to the square with number 64 moving from one square to another that it shares a side with having made no more than 14 moves. The greatest number of moves needed is when the numbers are in opposite corners of the board. If the numbers in any two adjacent squares differ by at most 4, then each move along our path will increase the number in the square by no more than 4. If the first square holds the number 1, then the second square will hold a number no greater than 5, the third a number no greater than 9, and so on. The last square will hold a number no greater than $1 + 14 \cdot 4 = 57$, yet the last square holds the number 64. This is a contradiction.

Solutions to Problem Set 22

Problem 22.1. Yes, it is true. If the number of mushrooms collected by Caroline is less than Anna's, but not the least, then she collected at least as many as one of the boys—perhaps at least as many as Dustin. Anna collected more than everyone else, so she collected more than Bryan. Therefore the girls collected more mushrooms than the boys.

Problem 22.2. What natural number could this be if its square is equal to $n(n+1)$? It can't be n, since its square is n^2, which is less than $n(n+1)$. Of course all numbers less than n do not work either. But $(n+1)$ also does not work because its square is $(n+1)(n+1)$, and is greater than $n(n+1)$. Similarly all numbers greater than $(n+1)$ do not work as the squares are too large. Therefore there is no such natural number whose square is equal to $n(n+1)$.

Problem 22.3. Remember that the plane can be colored in any way; any individual point can be colored one of the two given colors.

(a) Solution 1: Take a point of any color and construct a circle of radius 1 inch about it. If any point of the circle is the same color as the center, the center and that point are 1 inch apart and the same color. If all of the points on the circle are colored differently from the center, we can take any 1 inch long chord of the circle as our segment. The endpoints of the chord are of the same color and are 1 inch apart.

Solution 2: Examine any equilateral triangle in the plane with side 1 inch. All three of its vertices have been colored. Because there are three vertices and two colors, we will find at least two vertices of the same color. The distance between them will be 1 inch as required.

(b) This part is more complicated. Take two points with different colors— their existence is assumed in the problem—and label them A and B. For the sake of argument, let A be red with B green. If the distance between them is less than 2 inches, construct circles of radius 1 with centers at A and B. The circles will intersect in two points, and we will let C be one of the points of intersection. Then $AC = BC = 1$ inch, and point C is colored with one of two colors. If C is red, then points C and B are a pair we want. If C is green, then C and A is a suitable pair.

What can we do if $AB \geq 2$? Connect A and B with a segment, and divide it into pieces of length 1 by points $A_1, A_2, \ldots, A_n$, so that $AA_1 = 1$,

$A_1A_2 = 1$, and so on, until we reach a point A_n for which A_nB is less than 2 inches. What colors can the points $A_1, A_2, \ldots, A_n$ have? If one of them is green, take the first green one. The ones before it beginning with A are red, and now we've found a green one. This point and the one before it form the endpoints of a segment of length 1 inch with ends of different colors as desired. If all of the points $A_1, A_2, \ldots, A_n$ are red, then A_n and B are different colors and less than 2 inches from each other. In this case, we can find a segment of length 1 inch with ends of different colors as we argued in the previous paragraph.

Problem 22.4. This problem is similar to Problem 21.3.

(a) The answer is no. We can divide the 6×6 board into nine 2×2 squares. If the sum of the numbers in each of the 2×2 squares is negative, then the sum of all the numbers on the board will also be negative.

(b) Yes. Here is an example:

1	1	1	1	1
1	-4	1	-4	1
1	1	1	1	1
1	-4	1	-4	1
1	1	1	1	1

The sum of the numbers in each 2×2 square is -1, a negative number, while the sum of all of the numbers in the table is 5, a positive number.

Problem 22.5. Move n^2 from the left side of the equation to the right. We get an equation $m^2 - n^2 = 57$, or $(m-n)(m+n) = 57$. This equation expresses 57 as the product of two integer factors, $(m-n)$ and $(m+n)$. Since $(m+n)$ and 57 are positive, $(m-n)$ must be positive, and $(m-n)$ is certainly smaller than $(m+n)$. But 57 can be factored into natural number factors with the first factor less than the second one in only two ways: $1 \cdot 57$ and $3 \cdot 19$. Therefore there are two possibilities. One is $m - n = 1$, and $m + n = 57$ which gives $2m = 58$ so $m = 29$ and $n = 28$. The other is $m - n = 3$, and $m + n = 19$, which gives $2m = 22$, so $m = 11$ and $n = 8$.

Problem 22.6. First let's solve a slightly different problem. What is the greatest possible distance the car can travel with two pairs of tires? We will use the answer to solve the original problem.

Let's find a distance that is divisible by both 25000 miles and 15000 miles. For example, we can choose 75000 miles for this distance. Let's mentally drive 75000 miles in this car, changing tires as we need to. In going 75000 miles, we will use up 3 pairs of front tires and 5 pairs of rear tires for a total of 8 pairs. Therefore there is enough rubber on two pairs of tires to cover 1/4 the distance of 75000 miles, or 18750 miles. At what point should we rotate the front and rear tires so all tires will need to be replaced at 18750 miles? Driving half of 18750 miles, or 9375 miles, will consume half the usable rubber on two pairs of tires. Because the tires are identical and the appropriate rates of wear on each pair are constant, any excess loss of rubber on the rear tires is exactly compensated for by reduced wear on the front tires. Thus the front tires now have the same amount of useable rubber as the amount used up on the rear tires, and vice versa. If we now switch the tires, the car will be able to cover an additional 9375 miles before the other half of their rubber is used up. Thus both pairs of tires will be completely used up at 18750 miles if they are rotated 9375 miles from the start.

Problem 22.7. Let x and y be the given positive numbers. We will first show that $(x+y)^2 \geq 4xy$. Indeed, since $(x-y)^2 \geq 0$, we have $(x-y)^2 + 4xy \geq 4xy$; but the left-hand side of this inequality is $(x+y)^2$:

$$(x-y)^2 + 4xy = x^2 - 2xy + y^2 + 4xy = x^2 + 2xy + y^2 = (x+y)^2.$$

Next, since we are given that $xy > x+y$, we can write $(x+y)^2 > 4(x+y)$. Divide by $x+y$ to find that $x+y > 4$.

Problem 22.8. Number the boots, in order, from 1 to 30. Now examine three sets of 10 boots: the first 10; the next 10, and the last 10. There can't be more left than right boots in all three sets because then there would be more left boots than right boots, a contradiction. Similarly there can't be fewer left than right boots in all three sets. Therefore there will be a set of 10 with at least as many left boots than right boots, and there will be a set of 10 with no more left boots than right boots. If there is an equal number of left and right boots in any of these sets of 10, we have solved the problem. Otherwise let's move along the line of boots, one boot at a time, from a set of 10 with more lefts than rights to a set with fewer lefts than rights. If the set $\{1, 2, \ldots, 10\}$ has more left than right boots and the set $\{21, 22, \ldots, 30\}$ has are more right than left ones, examine consecutively the sets $\{1, 2, \ldots, 10\}$, $\{2, 3, \ldots, 11\}$, $\{3, 4, \ldots, 12\}$, and so on, up to $\{20, 21, \ldots, 29\}$ and finally $\{21, 22, \ldots, 30\}$.

For each set, record the number of left and right boots it contains, and compute the number of lefts minus the number of rights. In transitioning from one set to the next, we replace only one boot. Therefore, for all adjacent sets, the records will either not differ at all, or there will be 1 more of one type and 1 less of the other, in which case the difference will change by 2. The original difference is positive. This difference is also an even number because if there are x left boots in the first set of 10, then there are $(10-x)$ right ones there, and so the difference is $x - (10-x) = 2x - 10$, which is an even number. Therefore we start with a positive even number as the number of left boots minus the number of right boots, and, at each step, we either add or subtract 2. By the time we reach the last set of 10, the difference is negative. Therefore at some step we must have had a difference of 0. At this point we have a set of 10 consecutive boots with an equal number of left boots and right boots.

Problem 22.9. (a) Not always. As an example we can take skinny triangles: say, five right triangles each with hypotenuse 5 inches and one leg of length 1 inch. If we place four of them so that their short legs are horizontal and the fifth with its hypotenuse horizontal, we will not be able to cover the fifth triangle with the first four. We won't even be able to cover its hypotenuse, because each of the four others cannot cover a horizontal segment of length greater than 1 inch.

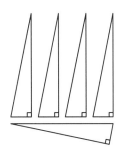

(b), (c). Yes, always. Suppose the sides of the original triangles are 2 inches long. Draw midlines in each triangle to divide it into four equilateral triangles with 1 inch sides. In the triangle we want to cover, circumscribe circles about each of its four triangles. In the remaining four triangles, circumscribe circles about the central triangle. All the circles will be congruent, and any circle around the central small triangle lies inside the original triangle. We can slide each of the four triangles so its central circle matches one of the four circles of the triangle we are covering. In this way each triangle we are sliding will cover one of the four little triangles in the triangle we are covering. We can cover four little triangles, and hence the big triangle, with four of the original triangles.

We have already solved (b). How can we use only three original triangles? Let ABC be the triangle we are covering. Let $L, M,$ and N be midpoints of sides $BC, AB,$ and AC, respectively. We will prove that circles circumscribed about triangles $MAN, MBL,$ and NCL already cover all of ABC, and therefore we don't need to use the fourth triangle to cover LMN. Let O be the center of triangle ABC. Drop perpendiculars from O onto the sides 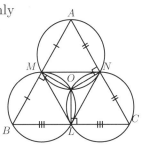 of ABC. They will end up exactly at the midpoints of its sides. Construct a circle with diameter AO. Because OMA and ONA are right angles, points M and N will lie on this circle. Therefore it coincides with the circle circumscribed about triangle MAN. This circle totally covers quadrilateral $OMAN$ which is exactly one-third of triangle ABC. Similarly, the circle circumscribed about MBL covers quadrilateral $OMBL$, and the circle circumscribed about NCL covers $ONCL$.

Problem 22.10. Divide the circle into five equal parts with five points. These points will be the vertices of a regular pentagon. Because there are five vertices and only two colors, we will find at least three vertices of the same color. They form a triangle we are searching for since any three vertices of a regular pentagon form an isosceles triangle. You should prove this last fact.

Problem 22.11. We will use Problem 21.2(c) several times. Specifically, we have the following inequalities:
$$a^2/b \geq 2a - b;$$
$$b^2/c \geq 2b - c;$$
$$c^2/d \geq 2c - d;$$
$$d^2/a \geq 2d - a.$$

By adding them all together, we get
$$\frac{a^2}{b} + \frac{b^2}{c} + \frac{c^2}{d} + \frac{d^2}{a} \geq 2(a+b+c+d) - (a+b+c+d) = a+b+c+d.$$

Solutions to Problem Set 23

Problem 23.1. Let ab (decimal notation, not a product) be the first two-digit number and cd the second. First, we find the number $4 \cdot ab$, then $4 \cdot ab + 7$, then $25 \cdot (4 \cdot ab + 7) = 100 \cdot ab + 175$, and finally $100 \cdot ab + 175 + cd + 125 = 100 \cdot ab + 300 + cd = (ab+3)100 + cd$. We can see that the last two digits of the resulting number form the second two-digit number. To find the first, we must remove the two last digits of the result and subtract 3 from the remaining number.

Problem 23.2. This is not possible. The line must pass through some vertex and cross the opposite side, for otherwise we will not get two triangles. The two angles formed in the new triangles by the intersection of one side of the original triangle with the line will sum up to 180 degrees, so one of them is at least 90 degrees, and hence one triangle will definitely not be acute.

Problem 23.3. No. Let's say John multiplied n by $(n+1)$, while Coleen multiplied m by $(m+2)$. Let's suppose that $n(n+1) = m(m+2)$. If $n \leq m$, the left-hand side of the equation will be at most $m(m+1)$, which is less than the right-hand side. If $n > m$, then $n \geq (m+1)$ because both are integers. Then the left-hand side at least $(m+1)(m+2)$, which is greater than the right-hand side. Therefore, there are no natural numbers n and m that satisfy the equation $n(n+1) = m(m+2)$.

Problem 23.4. Rewrite the given statements as follows: $x > 35$, $x < 100$, $x > 8 + \frac{1}{3}$, $x \geq 10$, $x > 5$. Notice that if $x \geq 10$, the third, fourth, and fifth inequalities are true, and either the first or the second one must also be true. But this contradicts the conditions in the problem. If $x \leq 8$, then the first, third, and fourth inequalities are false, which also contradicts the conditions. Therefore, $x < 10$, and $x > 8$, so the only possibility is $x = 9$, and it fits. The second, third, and fifth statements are true, while the first and fourth are false.

Problem 23.5. The desired point is the base of the height dropped onto hypotenuse AB from vertex C. Let M be a point on AB, and let MX and MY be the perpendiculars dropped onto the legs of the triangle from point M. We must locate M so that $MX^2 + MY^2$ is minimized. Notice that 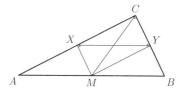 $MXCY$ is a rectangle, so

$MX^2 + MY^2 = XY^2 = CM^2$, since the diagonals of a rectangle are equal. Therefore, the sum $MX^2 + MY^2$ is minimized when the number CM^2 is minimized, which happens when the length of CM is minimized. The length of CM is, of course, minimized when CM is a height of triangle ABC.

Problem 23.6. (a) Draw a regular hexagon $ABCDEF$ in the plane. Suppose that the center O of this hexagon is white. If any two neighboring vertices of the hexagon, for example A and B, are also white, we have found an equilateral triangle AOB with all vertices of the same color. If every two neighboring vertices of the hexagon have different colors, the colors alternate. Then equilateral triangle ACE has all vertices of the same color. A similar argument applies if O is blue.

(b) Not necessarily. Suppose that the ink has produced alternating blue and white stripes, all of the same width. Also suppose that the lines separating the stripes alternate between blue and white, so one boundary of each stripe is blue and one is white, as depicted in the diagram. Further suppose that the width of the stripes is $\sqrt{3}/2$, the altitude of an equilateral triangle of side 1. Admittedly, that does not 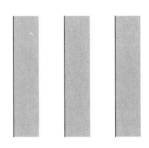 sound realistic, but the ink was magic and thus capable of producing any pattern!

Let's prove that if an equilateral triangle with sides 1 foot long has all vertices of the same color, then all its vertices must lie in one stripe. Clearly, they cannot lie in three different stripes of the same color, since otherwise the distance between some pair of vertices would be larger than 1 foot. Suppose that the vertices lie in two different stripes with vertices A and B in the first stripe, and vertex C in a second stripe of the same color (see first diagram below). These two stripes are separated by a middle stripe of a different color. Let M be the midpoint of AB. The point M must lie in the first stripe and, since the length of MC is equal to the width of the middle stripe, points A, B, and M can only lie on one boundary line of the middle stripe while point C must lie on its second boundary line, as in the second diagram below. But these lines are of different colors and therefore the points A, B, and C are not of the same color, which is a contradiction.

 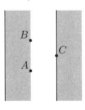

Thus all three vertices A, B, and C must lie in the same stripe. With an argument similar to the one above, we can show that in this case they must lie on two boundary lines of the stripe, and hence cannot be of the same color: again, a contradiction.

Problem 23.7. Both x and y must be greater than 2, for if $x \leq 2$,
$$\frac{1}{x} + \frac{1}{y} \geq \frac{1}{2} + \frac{1}{y} > \frac{1}{2}.$$

If $x = 3$, we find that $y = 6$, and if $x = 4$, we find that $y = 4$. Similarly, if $y = 3$, then $x = 6$, and if $y = 4$, then $x = 4$.

We have examined all of the cases where either x or y is at most 4. If both x and y are greater than 4, each of the fractions $1/x$ and $1/y$ is less than $\frac{1}{4}$, and their sum is less than $\frac{1}{2}$. Thus either $x = 3$ and $y = 6$, or $x = 6$ and $y = 3$, or $x = 4$ and $y = 4$.

Problem 23.8. Let's say it is now 9 a.m. of the day when the crab started descending the cliff to the beach. Imagine that another crab leaves the same beach and starts moving towards the top of the cliff, duplicating the way the original crab moved yesterday. Because both crabs are moving on the same path towards each other, they must meet. Their meeting point will be the point the first crab passed at the same time on the previous day as it did today.

Problem 23.9. It's clear that $c > a$ and $c > b$ because, otherwise, the left-hand side of the equation would be larger. Therefore, $(c-1) \geq a$ and $(c-1) \geq b$. Also, c is less than 3 because, otherwise, $c! \geq 3(c-1)! = (c-1)! + (c-1)! + (c-1)! \geq a! + b! + (c-1)! > a! + b!$. We must now examine the two remaining cases, $c = 1$ and $c = 2$. There are no solutions in the first case, while in the second we get $a = b = 1$. Therefore, $a = b = 1$, and $c = 2$.

Problem 23.10. Extend side BC to its intersection with side AD at the point M. Angles A and B of triangle AMB are 45 degrees each, and therefore this triangle is a right isosceles triangle. Similarly, triangle CMD is also a right isosceles triangle. Let $AM = x$ and $CM = y$. The area of the sail is equal to the sum of the areas of triangles AMB and CMD, so it is equal to $x^2/2 + y^2/2$. By the Pythagorean Theorem, $x^2 + y^2 = AC^2 = 4^2 = 16$. Therefore, the area of the sail is $8\,\text{m}^2$.

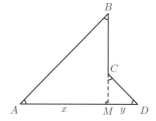

Solutions to Problem Set 24

Problem 24.1. The diagram shows the path of the center of the wheel. It consists of a series of arcs of a circle.

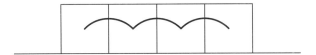

Problem 24.2. The answer is 4 mm. To understand this, you need to see how the books are arranged on the shelf. You can see from the diagram that the front cover of the first volume will be next to the back cover of the second volume.

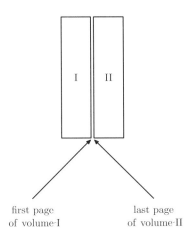

Problem 24.3. If there are x girls in the class, then there are $28 - x$ boys in the class. Let's represent the girls by x blue points, and the boys by $28-x$ green points. If a girl and a boy are friends, we will connect these two points with a line segment. Four segments emanate from each blue point, so there are $4x$ segments. On the other hand, three segments emanate from each green point, so the number of segments is $3(28 - x)$. Therefore we get the equation $4x = 3(28 - x)$, so $x = 12$. There are 12 girls and 16 boys in the class.

Problem 24.4. Solution 1: The speed of the minute hand is a one-minute division of the clock face per minute; the speed of the hour hand is one-twelfth of this speed. Suppose that the hands get together x minutes after 13:00. At 13:00 the minute hand will be five minute marks away from the hour hand. At x minutes after 13:00, the minute hand will be at the x minute mark on the clock face, and the hour hand will have gone through $x - 5$ minute marks. Thus $x = (x - 5) \div (1/12)$. Therefore $x = 12x - 60$, so $x = 60/11$. Since $60/11$ minutes will pass after 13:00, the hands will meet 1 hour and $60/11$ minutes after 12:00.

Solution 2: The hour hand moves at a rate of $1/12$ of the clock face per hour. The minute hand moves at a rate of 1 clock face per hour. Thus the hands get closer to each other at a rate of $1 - 1/12 = 11/12$ clock face per hour. So the next coincidence will happen in $12/11$ of an hour.

Problem 24.5. First we can find two points on the line, A and B, with the same color — let's say they are both green, with A to the left of B. Let C be the point to the left of A with $CA = AB$, and let D be the point to the right of B with $AB = BD$. If C is green, then CB is the segment we are searching for because A is its midpoint, and A, B and C are all green. Similarly if D is green, then AD is the desired segment. If C and D are both red, then we can look at the midpoint of AB, which is also the midpoint of CD. If it is green, then AB is the segment we are looking for; if it is red, then CD is.

Problem 24.6. Yes. Take any two gates, A and B. If they are connected by a walkway, we are done. Otherwise, at least seven walkways serve gate A, and at least seven more serve gate B. But these 14 walkways cannot all lead to different gates. If they did, there would be at least 16 different gates, including A and B, contrary to assumption. We conclude there is a gate connected directly to both A and B.

Problem 24.7. Solution 1: First let's examine the case when $x \leq y$. We can rule out $x \geq 3$, because then $xy \geq 3y = y + y + y > x + y$. Therefore x is either 1 or 2. If $x = 2$, then $2 + y = 2y$, so $y = 2$. If $x = 1$, then $1 + y = y$, which is impossible. Therefore $x = y = 2$. The case when $y \leq x$ can be examined in the same manner, with the same result.

Solution 2: Notice that

$$(x-1)(y-1) = xy - (x+y) + 1,$$

as you can check by expanding the product in parentheses. We see that if the equation $x + y = xy$ is true, then $(x-1)(y-1) = 1$ is also true, and vice versa. Therefore we have reduced the problem to finding all natural numbers x and y such that $(x-1)(y-1) = 1$. The product of the two nonnegative integers $(x-1)$ and $(y-1)$ is equal to 1 only when both of these numbers are equal to 1, so $x = y = 2$.

Problem 24.8. The areas of the semicircles in the diagram are equal to $\pi a^2/2, \pi b^2/2$, and $\pi c^2/2$. The area of the shaded region is then equal to

$$\frac{\pi a^2}{2} + \frac{\pi b^2}{2} + S - \frac{\pi c^2}{2} = \pi(a^2 + b^2 - c^2) + S.$$

By the Pythagorean Theorem, $a^2 + b^2 - c^2 = 0$, so the area of the shaded region is equal to S.

Problem 24.9. The good baron is right! Here are some examples:

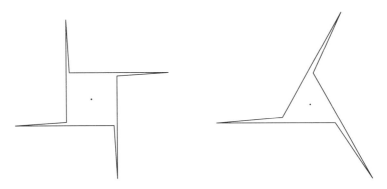

Problem 24.10. Let O be the point inside the square so that triangle COD is equilateral. From the diagram we can see that if we shift COD one unit to the left, it will coincide with triangle BKA. Because, under this shift, O moves to K, $OK = 1$. But $OC = OD = 1$ by the choice of O. Hence O 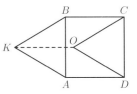 is equidistant from the vertices of triangle CKD, and therefore it is the center of the circle circumscribed about triangle CKD. The radius of the circumscribed circle is 1.

Problem 24.11. Draw two perpendicular blue lines on the plane. These lines divide the plane into four pieces. Number them clockwise 1, 2, 3, and 4. Color the interior of parts 1 and 3 blue, and color the interior of 2 and 4 yellow. Now recolor the point of intersection of the original lines red. We get a color scheme with every line colored with two colors, as you can check.

Problem 24.12. Number the rows of the board from 1 to 15. Do the same with the columns. Each rook now has two coordinates, the number of the row and number of the column it is in. Because the rooks cannot capture each other, they are on different rows and different columns. Because there are 15 rooks, there is exactly one rook in each row, and exactly one in each column. The sum of the row coordinates of all of the rooks will be $1 + 2 + \cdots + 15$, and the same is true for the column coordinates. Thus the total sum of all coordinates of all 15 rooks is an even number. When a rook takes a knight's move, it goes two squares in one direction and one square in a perpendicular direction, so the sum of its coordinates changes from odd to even or vice versa. By moving all 15 rooks in this way, we will change the sum of the coordinates from odd to even or even to odd 15 times. The sum was originally even, so it will be odd at the conclusion of all the moves. This would be impossible if the rooks were still unable to capture one another. Finally, if rook A can capture rook B, the rook B can capture rook A, and at least two rooks can capture each other.

Solutions to Problem Set 25

Problem 25.1. This is rather a question about conventions, since the answer depends on whether or not there was a year 0. Wikipedia informs us that neither the Gregorian nor the Julian calendar has the year 0. This means that the year 1 BCE was immediately followed by the year 1 CE. Hence the time period between May 1, 1 BCE, and May 1, 1 CE, is exactly 1 year long! Hence, the time period between May 1, 30 BCE, and May 1, 30 CE, is 59 years, and this is exactly how long the person in the problem lived.

Problem 25.2. Let us represent people by points. If two people know each other, we'll connect corresponding points by a green line segment, and if two people don't know each other, we'll connect those points by a red line segment.

(a) The problem can now be restated as follows: Five line segments emanate from a single point, each segment is either green or red. Prove that among these segments there are three of the same color. This is obvious: if, say, there are fewer than three green segments, then there are at least three red ones.

(b) This problem can be reduced to the following. Given six points, every two of which are connected by either a green or red line segment, prove that there exists a triangle with all three edges of the same color. Let us use part (a). Choose one of the points, call it A. Five segments emanate from A, so we can choose three of these segments which are of the same color, say red. Suppose these red segments lead to points B, C, and D. If any two of these points are connected by a red segment, then these two points together with point A form a triangle with all red edges. Otherwise, all points B, C, and D are connected by green edges and hence they form a triangle with all green edges.

Problem 25.3. Solution 1: Since the average age is 22, the sum of the ages of all players is $11 \times 22 = 242$ years. After the injured player left, the sum of the ages of the remaining players is $10 \times 21 = 210$ years. Thus the age of the injured player is $242 - 210 = 32$ years.

Solution 2: By distributing years evenly, we get 22 years per player. After the injured player left, we now get 21 years per player. Hence the injured player took out "his" 22 years, plus 1 year from each of the 10 remaining players, so altogether we get 32 years.

Problem 25.4. To begin with, the cat sits in one room and cannot reach any of the remaining 8 rooms because all doors are closed. Let us mark those doors that we want to open, and start opening them in sequence so that each new open door leads from a room that the cat could already reach. Clearly, at each step we add no more than one new room. Thus we need to open at least 8 doors.

On your own, you should find 8 doors to open which would allow the cat to freely roam through all 9 rooms of the apartment.

Problem 25.5. (a) Every time the tourist visits Clearview Island, he needs to traverse two bridges — one bridge to come to the island, and another one to leave it. Thus there are six bridges on Clearview Island.

(b) The tourist left Clearview Island by one bridge, then he used two other bridges every additional time he visited the island. Thus we have a total of five bridges.

(c) The tourist used one bridge to leave the island in the beginning, and he used one more to come back at the end. He used two more bridges when he visited the island in the middle of his journey. Thus we have a total of four bridges.

Problem 25.6. We have already solved a similar problem. Suppose that the table has m rows and n columns. Let's find the sum of all the numbers in the table in two ways: first by rows, then by columns. One way we get $19m$, and the other way we get $19n$. Therefore $19m = 19n$, so $m = n$.

Problem 25.7. Let's make a very important observation. If the bug has neither started nor ended at a certain vertex, then the total number of edges emanating from the vertex must be even. This is so because the bug must crawl along exactly two edges through such a vertex every time he visits the vertex. Thus only the starting and ending vertices can have odd number of edges coming from them. But a cube has 8 vertices, and each of these vertices has an odd number of edges. Thus the bug won't be able to crawl through all the vertices as required.

Problem 25.8. Yes, it is possible. We can use the solution of Problem 24.9. In the diagram to the right, we took the point and the polygon from that problem, made a small incision on one of the sides, and connected its ends by a broken line in such a way that the interior of the old polygon became part of the exterior of the new one. The chosen point lies outside of the new polygon and none of its edges is totally visible from this point.

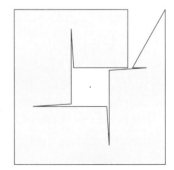

Set 25   181

Problem 25.9. (a) Suppose the first student is Alice. To begin with, students numbered $2, 4, \ldots, 16$ will sit down. We now get a circle with eight students, and Alice is again the first. We'll again move around the circle, "crossing out" every other student, and will obtain a new circle consisting of four students, where again Alice is the first one, and so on. Thus the last remaining student is Alice, the first one. This conclusion will be the same if the number of students is any power of two. Each time around the circle we'll reduce the number of students by half, and we'll start counting again from the first student. This student will be the last remaining one.

(b) After student number 2 sits down, there will be 16 students left in the circle, and we will start counting from the third student. Now the problem becomes exactly as part (a) above: the last remaining student is student number 3 in the original circle of 17.

(c) After students number 2 and 4 leave, there are 32 students left, and the count starts with the student number 5. Since 32 is a power of two, we see, using the argument of part (a), that the last remaining student will the one we are now counting on, namely student number 5.

(d) The largest power of 2 which does not exceed 1000 is $2^9 = 512$. In order to get 512 students in the circle, 488 students must leave. Since every other student gets out, the last one who leaves will have the number of $2 \times 488 = 976$, after which the count starts with student number 977, and the number of the remaining students is a power of 2. By part (a), the last remaining student will be student number 977.

Problem 25.10. The total number of the games played was at least 21 because that is the number Victor played. Notice that no one can skip two games in a row; if a person skips one game, they will play the next one with the winner of the previous game. If the total number of games were 22 or more, then Ivan would have played at least 11 games because he plays at least every other game. But Ivan played only 10 games, so exactly 21 games were played. But then Victor played in all these games, and each of these games he played either with Ivan or with Ashley. Since he played Ivan 10 times, he played Ashley 11 times. Thus Ashley played 11 times, and all her games were against Victor.

Problem 25.11. The answer is yes, it's true. Let x be the number of wallets, and let y be the number of pockets. Since the number of dollars in every pocket is less than x, the number of wallets, the total number of dollars in all pockets is less than xy. Suppose that the number of dollars in every wallet is at least y, the number of pockets. Then the total number of dollars in all the wallets is at least xy, a contradiction. Thus there exists a wallet containing a number of dollars that is less than the number of pockets.

Solutions to Problem Set 26

Problem 26.1. Carts are getting closer to each other with the speed of $10 + 15 = 25$ km/h, and the distance between them is 50 km when the fly starts flying, so they will meet two hours after the fly starts flying. Because the fly's speed is 20 km/s, the fly will cover 40 km in those two hours.

Problem 26.2. The length of the wire is exactly equal to the total length of the edges of the cube, so each edge will have a single strand of wire along it. Let's imaging building the cube out of the minimal number of pieces of wire. As we add wire pieces, we can keep track of the number of edges coming out of each vertex of the cube. Where the wire is bent somewhere in the middle to go through a vertex, that vertex will get two additional edges coming out of it. To form a cube, each vertex must have three edges coming out of it. It is only where the ends of a piece of wire connect to a vertex that the parity of the number of edges at that vertex can change from even to odd. Each piece of wire has just two ends, so each additional piece of wire can change the parity of the number of edges for at most two vertices. The cube has 8 vertices, so we will need at least $8 \div 2 = 4$ pieces of wire.

See if you can figure out how to build a cube having cut the wire into four pieces of the appropriate lengths.

Problem 26.3. We have already solved a similar problem. We can take the length of a side of one of the grid squares as being of unit length. Suppose that the side of the big square is x units, and the side of the hole is y units. Then the remaining figure consists of $x^2 - y^2$ grid squares.

(a) We need to solve the equation $x^2 - y^2 = 9$, or equivalently, $(x+y)(x-y) = 9$. Since x and y are natural numbers and $y < x$, we must have $x + y = 9$ and $x - y = 1$. Therefore $x = 5$ and $y = 4$.

(b) Here we need to solve the equation $x^2 - y^2 = 10$ or $(x+y)(x-y) = 10$. The numbers $x + y$ and $x - y$ differ by $2y$, and therefore are of the same parity. They can't both be odd since the product of two odd numbers can't be 10. But they can't both be even either, since the product of two even numbers is divisible by 4 while 10 isn't. Thus it is impossible to have 10 grid squares remaining.

Problem 26.4. (a) Let us leave one of the towns and wander along the roads while observing the rule that we cannot traverse the same road more than once. Clearly, we won't be able to travel indefinitely, and thus we will

Set 26   183

eventually get to a town from which we cannot go farther. It cannot be a town that we have already visited because if this were so, we would have made a loop during our travel. It would then be possible to get between any pair of towns in the loop in two different ways by going in opposite directions around the loop. This contradicts the assumption that there is only one route between any two towns. Thus we will get to that last town by the single road leading to it. This is a town with only one road out of it.

Having found one town with exactly one road emanating from it, let's start a trip from this town while observing the rule that we cannot traverse the same road twice. As before, we will eventually reach a dead end. This is a second town with exactly one road out of it.

(b) We can use the results of part (a). We have found a town with exactly one road coming out of it. Let's mentally remove it together with the road. We're left with a country where it's possible to get from any town to every other one in a unique way without using any of the roads more than once. We can now find a town with exactly one road coming out of it, we can remove this town and road, and so forth. By continuing in the same fashion, we will eventually remove $N-1$ towns and $N-1$ roads, and there will be one town and no roads remaining. Thus there are $N-1$ roads in the country.

(c) No. Let A and B be the two towns at the ends of the road that was closed for repairs. If it is still possible to get from A to B, then to begin with, there would have been two different ways to get from A to B — the other way was by the closed road. But this contradicts the uniqueness condition.

Problem 26.5. Suppose the distance between the villages is x miles. Those stretches that are uphill going in one direction become downhill going the opposite way, and vice versa. Thus during a round trip the bus goes x miles downhill and x miles uphill, and hence it takes $x/15 + x/30$ hours. Thus $x/15 + x/30 = 4$, so $2x + x = 4 \times 30$, so $x = 40$. The distance between the villages is 40 miles.

Problem 26.6. (a) Let's take three identical balls and place them on a table so that every two of them touch. The points where the balls touch the table form an equilateral triangle with side twice the radius of a ball. Now put a fourth ball of the same size on top of the three balls so that it touches each of them. As a result the centers of the four balls will form a regular tetrahedron with side equal to twice the radius of a ball.

(b) Take the pyramid of four balls that we constructed in part (a). From the center of this pyramid, let's start expanding a little sphere until it touches all four of the original balls, which it must do by the symmetry of the pyramid.

Problem 26.7. If we cut all horizontal ropes except the topmost one in each column, we will have cut $50 \times 600 = 30\,000$ ropes, and the net will remain connected.

Let us prove that it is impossible to cut more ropes. Imagine that the vertices of the squares are towns and the ropes are roads. If we have cut as many ropes as possible, it would mean that it is impossible to get from one town to another in more than one way while using each road at most once. Otherwise, there would be a loop, and we could remove one of the roads. Thus, in terms of the net, we'd be able to cut one more rope without disconnecting the net. We're in the same situation as in Problem 26.4. The number of towns here is the same as the number of vertices of the squares which is $51 \times 601 = 30\,651$. According to Problem 26.4(b), the number of ropes left uncut is $30\,650$. But the total number of ropes initially was $51 \times 600 + 50 \times 601 = 60\,650$, so the maximum number of ropes which could be cut is $30\,000$.

Problem 26.8. Simultaneously light one of the fuses on one end and the other on both ends. The second fuse will burn down in half an hour, and, at this moment, light the other end of the first fuse. From that moment, the fuse will burn down in 15 minutes. Thus, from the start to the moment the first fuse burns down, exactly 45 minutes have elapsed.

Problem 26.9. On the diagram below, we have marked congruent angles with the same digits. The angles are congruent because of the symmetry of a circle. The sum we are looking for is equal to the sum of the angles marked by 1, 2, and 3 at the point where all three circles intersect. Obviously, this sum equals 180 degrees.

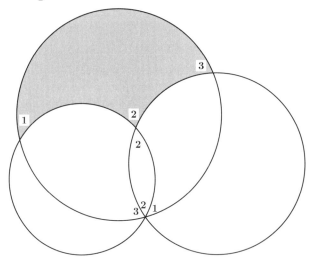

Problem 26.10. Let's choose one of the 17 points and call it A. There are 16 colored segments coming out from A. At least six of these segments are of the same color — say they are all yellow. These six yellow segments connect A with six other points. If at least two of these six points are connected by a yellow segment, these two points together with A form a yellow triangle. If none of the segments connecting these six points is yellow, then we have six points connected by green and red segments. This is the situation in Problem 25.2(b), and we can find a monochromatic triangle in this case.

Solutions to Problem Set 27

Problem 27.1. This problem is a bit of a joke: The idea is that there are 365 days in an ordinary, nonleap, year. There are 12 months in a year: one of these (February) has 28 days; four (April, June, September, and November) have 30 days each; and the remaining seven have 31 days each. Therefore for $x = 1, y = 4, z = 7$ the equation is true.

Problem 27.2. Andy shot a total of 25 times. Exactly 20 of them were given to him for hits. At two shots per hit, Andy hit $20 \div 2 = 10$ times.

Problem 27.3. The smartest student reasoned as follows: I see two kids with dirty faces. If my face is clean they would not laugh at me but at one another. But then the smarter of the two would soon realize that his face is also dirty, and he would stop laughing. Because they continue to laugh, my face is dirty too.

Problem 27.4. (a) Yes it can, for example, $5544 + 4455 = 9999$.

(b) No, it can't. Let's assume the opposite. Because the number of digits in both numbers is identical, they must contain five digits each or the sum would be smaller or larger than 99 999. Also, while adding these numbers, we can't carry over digits. The sum of any two digits is no greater than 18, so adding the rightmost digits of the numbers, we get exactly 9 with no carryover. Similarly, adding the second to last digits we get exactly 9, and so on. But then the digits of our numbers can be paired up so that each pair has the sum of 9. Because one of the numbers is formed by transposing the digits of the other, the digits of this single number can be split into pairs that sum up to 9. But this is impossible, since the number has five digits, and there is no digit that could form a pair with itself because the sum of any digit with itself is an even number, while 9 is odd.

Problem 27.5. (a) It turns out that it's enough to cut only one link—the third from an end. Then the chain divides into three pieces consisting of one link, two links, and four links. On the first day, the traveler gives the innkeeper one link. On the second day, he gives two links and gets one link back. On the third day he gives one link. On the fourth day, he gives four links and get both the one- and two-link pieces back. In the remaining three days, the traveler will repeat what he has done in the first three days, only now the innkeeper keeps the four-link piece.

(b) It's enough to cut only two links—the fourth and the eleventh from an end. The chain will then be divided into five pieces consisting of one link, one link, three links, six links, and twelve links. You can check yourself that using these pieces it is possible to pay the innkeeper by his rules. Now prove that cutting only one link is not enough.

Problem 27.6. Let's mark all the squares where a knight can get, allowing several moves, by placing a dot in the center of such a square. Let's number these dots, and connect two dots if a knight can get from one marked square to another after exactly one move. We obtain the top picture shown on the right.

To begin with, the knights are at the four given dots, and a single knight's move corresponds to displacement from one dot to another along a connecting segment. We can observe that the dots and segments form a cycle. This is easier to see if we draw the numbered dots and connecting segments without the board. The colors of knights in the original arrangement alternate, as seen in the middle diagram on the right. While knights are going around in a cycle, the colors of the knights will always alternate since a knight cannot pass another knight in the cycle. In the target arrangement, shown in the bottom diagram, the colors do not alternate, so it is impossible to place the knights as desired.

Problem 27.7. The sequence of steps is shown below. We first place one chain on a table, then place a 2×3 rectangle on top of it. Next we place the second 2×3 rectangle on top of the first, but turned 90 degrees, as in the third diagram. Finally, put the remaining chain on top of this rectangle, also perpendicular to the first chain. Our pyramid is finished! You should try to make the construction yourself (maybe with gum drops—yum!) to check the solution.

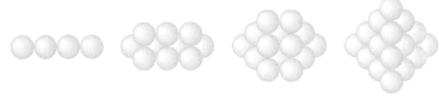

Problem 27.8. Buckets made as truncated cones can be stacked more compactly for convenient storage and shipping. A similar principle is used for plastic cups and supermarket carts.

Problem 27.9. (a) If $S = 1$, there is only the surrounding sea; therefore the dams can't form a loop because the loop would enclose a second sea. With no loops, it is impossible to have two different ways to get from one island to any other one if we move along the dams and use every dam at most once. Thus we find ourselves in the situation of Problem 26.4, except that here the islands correspond to towns, and the dams correspond to roads. Thus, by Problem 26.4(b), $D = I - 1$, so $I - D = 1$. Adding $S = 1$ to the previous equation, we obtain $I - D + S = 2$ as desired.

(b) If the number of seas is greater than one, then there are two seas which are adjacent to each other along a dam. Let's destroy this dam. Notice that we can still get from any island to any other one using the remaining dams. Indeed, we have destroyed a dam which separated two seas, and one of these seas was not the big outer sea, and thus it was bounded by a closed path consisting of dams. We have destroyed one dam in this loop, but the islands directly connected by that dam are still connected by the remaining dams in the loop.

(c) If there is only one sea, we're done by part (a). Otherwise, destroy a dam as we did in part (b). We obtain exactly the same problem, only the number of seas and dams decreases by one each. Again, if the number of seas is still bigger than one, destroy a dam as in part (b), and so on. Eventually, having destroyed n dams, we will come to the situation with exactly one sea, as in part (a). The number of islands has not changed; the number of dams is now $D - n$, and the number of seas is $S - n$. From the formula in part (a), we now have
$$I - (D - n) + (S - n) = 2,$$
so $I - D + S = 2$, as desired.

Problem 27.10. Let's divide the entire board, except for the central square, into 2×1 dominos. For example, it can be done as shown in the diagram. At each move, Ann shifts the marker into some square of some domino. If Beth then moves the marker to the second square of the same domino, she will always be able to move, and thus Beth will win.

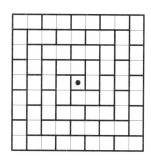

Solutions to Problem Set 28

Problem 28.1. First fill the pot with water, then dump the extra water by tilting the pot until the bottom just starts to show, as in the diagram on the right.

Problem 28.2. The diagonal of a square is longer than its side. Thus, if carelessly handled, a square cover can easily be dropped into a square manhole, as shown on the left (thick lines indicate the hole).

Problem 28.3. An imprint will only occur in the right top square of the figure, and the letter P will be turned 90 degrees clockwise, as shown on the right.

Problem 28.4. The answer is yes. Place three pencils on a table so that two of them touch at one end and a third is wedged in between. Then place three more on top of these, in the same configuration but rotated 90 degrees, as shown on the right.

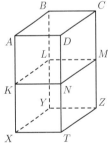

Problem 28.5. It should be clear from the way the problem is posed that the aquarium is a $1 \times 1 \times 2$ rectangular parallelepiped, and this is reflected in diagram on the left. Let's assume that the fish swims from the top down.

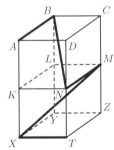

(a) From the leftmost diagram of the statement of the problem, we see that the fish started at a point on edge AB; the right rectangle then shows that it swam first from A to B along edge AB. Then we see that it swam directly along a straight line to point N, then along the line segment NM, then straight on to the point X, and finally along edge XT to the point T. The entire path of the fish is shown on the right.

The desired answer (the view from above) is this:

Set 28   189

(b) It can be seen from the diagrams in the statement of
the problem that the fish started at point B and swam
along edge BL. Then it went to point N, but it did not
necessarily swim along a straight line. It could swim
along any curve as long as it remained in the plane
$KLMN$. From point N, the fish swam straight to point
X, and finally it swam along edge XT to point T. One
possible path for the fish is shown on the right.

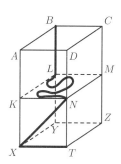

The view from above could be something like this:

Problem 28.6. No, since the openings of the two bottles are the same size!

Problem 28.7. Only the leftmost cube. You should be able to prove it.

Problem 28.8. Yes it can. Let's place the cube so that one of its diagonals — a line connecting two most distant vertices of the cube — is perpendicular to the ground, and let light rays parallel to the diagonal shine on the cube from above.

Problem 28.9. Let's place the bricks as shown
in the diagram, with two bricks stacked up and
the third one placed side by side with the bottom
brick in the stack. In this configuration, there is an empty space above
the third brick, with the same dimensions as each of the bricks. The two
marked points on the diagram represent opposite vertices of this imaginary
brick, and there are no obstacles to measuring the distance between these
two points with a ruler. This distance is equal to the length of the diagonal.

Problem 28.10. (a) Yes, it is possible. For example, we can take a cube
and glue an identical cube to each of its faces.

(b) No, it is impossible. At every vertex of such a polyhedron, there are
exactly three faces — in fact it's impossible to have fewer than three faces
at a vertex of any polyhedron — while in the convex case it's impossible to
have more than three square faces since otherwise the sum of the flat angles
at the vertex would be at least $4 \times 90° = 360°$, which is impossible for a
convex polyhedron.

Problem 28.11. Yes, *Alpha* will be able to sail first without disturbing
Kvant's mooring cable. How to do this is shown below. The author of the
problem, Nikolay Dolbilin, actually witnessed a similar situation at a port.
At the last step, considerable force was required to jerk the cable out, so a
dock worker used a winch.

Problem 28.12. Let's solve part (b) right away. Consider a regular tetrahedron placed in space in such a way that our bulb is exactly at its center. Circumscribe a circle around one of the faces of the tetrahedron. Connect the bulb to every point of this circle with rays emanating from the bulb. We have constructed an infinite cone. We can inscribe a ball of any radius in this cone, and it will block all light rays from the bulb that are inside this cone. But what about the rays along the cone's surface? Are they blocked by the ball? To be on a safe side, let's slightly increase the size of the ball without changing the position of its center while making sure the new ball still does not touch the bulb. Now the rays along the cone's surface will also be blocked. In a similar manner, consider a second cone gotten by using a different face of the tetrahedron. Rays in this cone can be blocked by a ball of a suitable radius placed far enough from the bulb so that it does not intersect the first ball. Similarly, we can construct and block two more cones. Since every ray emanating from the bulb lies in one of the four cones, the four balls constructed in this way block all the light emitted by the bulb.

Problem 28.13. Let's draw a circle on the ball and call it α. To do so we use the compass (see the diagram on the right). Open the compass a distance a, and place the sharp point of the compass at a point A on the ball. What is the radius of α? At first glance it might appear that the radius of α is a, but this is not so. Where does the center of the circle α lie?

Let's mentally cut the ball with a plane containing α. This cut will separate off a "cap" of the ball, while the cross-section itself will be a disc bounded by α, whose center, S, will be the center of α. The straight line AS is perpendicular to the plane containing α, and it passes through the center, O, of the ball. If we mentally cut the ball into two halves with a plane containing the line AS, we get the picture on the right. In this picture the circle α appears as the line segment CD. The lengths of AC and AD are equal to a. If we knew the radius of α, i.e. the length of CS, we would be able to construct triangle CAD. But then we would also be able to find the center O of the circle circumscribed about triangle CAD — it is the intersection of the perpendicular bisectors of the sides of CAD — and thus we'd be able to find the radius of the ball which is equal to the length of OA.

But how can we find the radius of α? The idea is to construct on paper a triangle whose circumscribed circle is congruent to α. Draw a new circle, β, on the ball by placing the sharp point of the compass at some point, B, on the circle α and using the same compass opening, a (see drawing immediately above).

This new circle will pass through A and will intersect α at two points X and Y. Clearly, $BX = BY = a$. Place the compass's legs at points X and Y, and, without changing the opening of the compass, mark this distance on paper. This gives the picture in the last diagram, where the points X_1 and Y_1 are such that $X_1 Y_1 = XY$. Now, with the opening of the compass back to a, con- 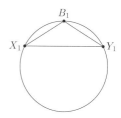 struct on the paper a point B_1 such that $B_1 X_1 = B_1 Y_1 = a$. Triangle $X_1 B_1 Y_1$ is congruent to triangle XBY. Now we can construct the circle circumscribed about triangle $X_1 B_1 Y_1$ on the paper, and its radius is equal to the radius of α.

Note: For a plane, the following amusing fact is true. Let's draw a circle on a plane, mark on the circle a point, and start marking off points on the circle one by one without changing the compass's opening: first place the sharp point of the compass at the first marked point and mark the second one, then move the sharp point to the second mark and mark the third point, and so on. At the sixth step, we'll come back to the first point; moreover, the six marked points are the vertices of a regular hexagon. If we try doing the same thing on a sphere, we won't get a regular hexagon (think about the reason). So if you find yourself on an unknown surface, you can carry out this experiment—if you end up with something other than a regular hexagon, you can conclude that you are not on a plane!

Mathematical Maze

The Mathematical Maze is a fun and lively competition. Several rooms are prepared beforehand, each containing a number of problems on a certain subject. Students are given the map of the "maze" and are told the rules. The map has room numbers and names indicating the kind of problems they contain: examples of six "rooms" are given in the next few pages.

Each student is supposed to visit all the rooms in any order, and solve one problem in each room. At the end the map will have time stamps for entering and exiting the room. It will also have the number of the problem solved in that particular room.

The main goal of the game is to visit all the rooms. The rooms can be far apart, sometimes even on different floors. Some students race between rooms at top speed, startling their parents who are waiting in the corridors. The game is always fun, giving kids a chance to both think and run around.

It may happen that all problems in a certain room are too difficult for a particular student to solve. Then that student is sent to a special room, called "Intensive Care", where he gets a similar but easier problem. The student receives credit for the problem, but loses time.

A student who makes a successful visit to all the rooms on the map goes to the prize auditorium, where she receives as a present a book of her choice, from a list prepared by the instructors. Although there are enough books for all the students, the sooner a student finishes the maze, the more choices she has. If a student comes to the prize auditorium very early, he may be offered an interesting extra problem as a bonus. Some students ask permission to go through the maze a second time, and some even manage to do it in time to get another book.

Mental Arithmetic

In this room, students are given a problem and are allowed to spend some time thinking about it. They then have to give the answer out loud. They have to solve the problem in their heads, without writing.

MA.1. Replace ∗ with the sign of an arithmetic operation in order to get a true equation:
$$0.375 * \frac{1}{40} = 0.4.$$

MA.2. What is 17 multiplied by 34?

MA.3. Replace the stars with digits in the number 52∗2∗ so that the result is divisible by 36.

MA.4. Three brothers had a total of 18 pencils. The youngest brother had two fewer pencils than the middle brother, while the oldest had two pencils more than the middle one. How many pencils did each brother have?

MA.5. Replace the stars with numbers in order to get a correct equation:
$$\frac{9}{*} - \frac{*}{21} = \frac{17}{42}.$$

MA.6. What time is it if the time remaining in the 24-hour day is twice what has already passed? The day starts at 12:00 am.

MA.7. Eight 8's are written in a row: 88888888. Place an addition sign between some of them so that the resulting sum equals 1000.

MA.8. How many times should the largest two-digit number be added to the largest one-digit number in order to get the largest three-digit number?

MA.9. A boy has as many sisters as brothers, while his sister has twice as many brothers as she has sisters. How many brothers and sisters are in this family?

MA.10. What is 43 multiplied by 17?

MA.11. There is total of 30 notebooks in two stacks. If two notebooks from the first stack are moved to the second stack, then the first stack will contain twice as many notebooks as the second one. How many notebooks are there in each stack?

MA.12. Replace the ∗ with the sign of an arithmetic operation to get a correct equation:
$$37.3 * \frac{1}{2} = 74\frac{3}{5}.$$

MA.13. Susan opens a book at random and tells Alex: "Guess what page I'm looking at. If you add 5 to the page number, then divide the result by 3, then multiply by 4, then subtract 6, and then divide by 7, the result is 2." What answer should Alex give?

MENTAL ARITHMETIC

MA.14. The American painter and sculptor Jonathan Borofsky (born in 1942) filled thousands of sheets of paper with consecutive handwritten numbers, starting from 1. How many numbers had he written by the time he put down the 1392nd digit?

MA.15. Which is greater, $\frac{1}{2} - \frac{2}{3} + \frac{3}{4}$ or $\frac{3}{6} - \frac{8}{12} + \frac{15}{25}$?

MA.16. Alan is now 11 years old and Bob is 1 year old. How old will Alan and Bob be when Alan is three times as old as Bob?

MA.17. Place appropriate parentheses in
$$4 \cdot 12 + 18 \div 6 + 3,$$
so that the result is equal to 50.

MA.18. Find two numbers whose sum is 12 and are such that if each number is squared, the sum of the two squares is 80.

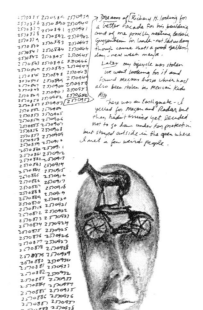

Jonathan Borofsky, "Page from my Notebook", ink and crayon on paper, 1978. Used by permission.

MA.19. Pat has three brothers. One is three years older than Pat, another is three years younger than Pat, and the third has one-third of Pat's age. Their father has three times Pat's age, and the sum of all their ages is 95. How old is Pat?

MA.20. Replace both asterisks with natural numbers so as to get a correct equation:

$$\frac{5}{*} - \frac{*}{3} = \frac{1}{6}.$$

MA.21. This problem is from ancient Babylon and dates to arround 2000 BCE. The length and a quarter of the width together measure seven handspans, while the length and the width measure ten handspans. How many handspans are there in the length, and how many are in the width?

MA.22. Twenty 5's are written in a row: 55555555555555555555. Place addition signs in between some of them so that the resulting sum equals 1000.

MA.23. Which of the following two numbers is larger: $12 \cdot 7 \cdot 30 \cdot 2$ or $3 \cdot 14 \cdot 15 \cdot 8$?

MA.24. Sally has thought of a number. The sum of one third of the number and one quarter of the number is 21. What is the number?

MA.25. What is 27 multiplied by 44?

MA.26. If Igor bought three notebooks, he would still have 11000 rubles left, but if he wanted to buy nine of these notebooks he would be 7000 rubles short. How much money does he have?

MA.27. Replace the ∗ with the sign of an arithmetic operation in order to get a correct equation:
$$\frac{33}{40} * \frac{10}{11} = 0.75.$$

MA.28. Jason borrowed a book from a friend for three days. On the first day he read a half of the book; on the second day, a third of the remaining pages. On the third day the number of pages that he read was equal to one half of the number of pages he read during the first two days. Has Jason finished the book?

MA.29. How should one place plus signs among the digits 1234567 in order to get an expression whose value is 100?

MA.30. It takes ten days for Pirate John
To gobble up a cask of rum.
The same rum lasts for Pirate Zeke
A trifle longer: two whole weeks.
How long will it take for the rum to end,
If together their time they choose to spend?

MA.31. Pierre is paid 48 francs for each day he works, and he is fined 12 francs for every work day he misses. At the end of six work weeks (30 days), Pierre ends up making no money at all. How many of these 30 days did he work?

MA.32. Replace the asterisk with an operation sign so that the equation is correct:
$$0.45 * \frac{1}{20} = \frac{2}{5}.$$

Two-Player Games

The goal of these game problems is to try to figure out which player — the first to make a move or the second — can guarantee a victory, regardless of how the opponent plays, and how he should play to do so.

A student is assigned a game and given time to consider what strategy to use. The student then plays the game with the instructor twice: once playing first, the second time playing second. A victory in either attempt earns the student full credit.

TWO-PLAYER GAMES

TPG.1 (New Years' Eve). Two players take turns moving the hour hand on a 12-hour clock either two or three hours forward. The hour hand starts at 6, and the winner is the player who can make the hour hand point to 12.

TPG.2 (Cross Out a Row). On a 9×10 grid, two players take turns crossing out either a row or a column as long as it contains at least one square that has not been crossed out. The winner is the last player able to make a move.

TPG.3 (Race to 100). The first player names any whole number from 1 through 9. The second adds any whole number from 1 through 9, to the first number. The players alternate turns adding a number from 1 through 9. The winner is the first player to reach 100.

TPG.4 (Eight Signs). Eight minuses are written in a row. Two players take turns changing either one minus or two adjacent minuses to pluses. The player who changes the last minus to a plus wins.

TPG.5 (Get 0). The game starts from 60. Players take turns decreasing the current number by any of its divisors. The player who gets to 0 loses.

TPG.6 (Tic-Tac-Toe). Players take turns placing X's and O's in a 3×3 grid with the first placing X's and the second placing O's. Each player's goal is to make a complete row—horizontal, vertical, or diagonal—with the same symbol.

TPG.7 (Nautical Wheel). A ship's wheel has eight handles. Two players take turns knocking off either one handle or two adjacent handles from the wheel. Whoever knocks off the last handle wins.

TPG.8 (Points on a Circle). Given ten points on a circle, two players take turns drawing straight line segments between pairs of those points. A player may not draw a segment that meets another segment in the interior of the circle. A segment that meets another segment only in one of the original points is alright. The last player to be able to draw a line segment wins.

TPG.9 (Nine Signs). Nine minuses are written in a row. Two players take turns changing either one minus or two adjacent minuses into pluses. The player who changes the last minus into a plus wins.

TPG.10 (Loves Me—Loves Me Not). A daisy has nine petals. Two players take turns picking either one petal or two adjacent petals from the daisy. The winner is the one who removes the last petal.

TPG.11 (King in the Corner). A king is in the upper left corner of a chessboard. Two players take turns moving the king. The king can move one square at a time either to the right, down, or

down and to the right diagonally. The winner is the player who can put the king in the bottom-right corner.

TPG.12 (Pebbles). Two players take turns taking either one pebble from one of two piles of 8 pebbles, or one pebble from each of the two piles. The winner is the one to take the last pebble.

TPG.13 (Mutant Tic-Tac-Toe). Two players take turns placing X's and O's on a 7×7 grid. The winner is the player who can place three of his symbols in a row either horizontally, vertically, or diagonally.

Geometry

G.1. Prove that a circle constructed with one side of an equilateral triangle as a diameter passes through the midpoints of the two other sides.

G.2. Prove that a circle constructed with one of two equal sides of an isosceles triangle as a diameter passes through the midpoint of the third side.

G.3. Suppose that a circle constructed with the side of a triangle as a diameter passes through the midpoint of another side. Prove that the triangle is isosceles.

G.4. Suppose two altitudes of a triangle are equal. Prove that the triangle is isosceles.

G.5. Suppose the bisectors of two angles of a triangle form a 120 degree angle when they intersect. Find the third angle in the triangle.

G.6. Suppose angle A in triangle ABC is obtuse, and the perpendiculars to AB and AC through A divide BC into three equal parts. Find the angles of ABC.

G.7. Show how to construct a triangle using a compass and a straightedge if you are given one of its sides, one of the two angles adjacent to the side, and the sum of the remaining two sides.

G.8. The bisectors of angles B and C in triangle ABC intersect at point K. The line parallel to BC passing through K intersects AB at point M, and AC at point N. Prove that $MN = BM + CN$.

G.9. Points K and M have been chosen from the hypotenuse AB of a right triangle ABC, with $AK = AC$ and $BM = BC$. Find angle MCK.

G.10. Angle A in triangle ABC is equal to 60 degrees. Prove that $AB + AC < 2BC$.

G.11. Given a circle with its center and a point A outside the circle, use a compass and straightedge to construct a line tangent to the circle that passes through A.

G.12. A height and a median from the same vertex of a triangle divide that angle into three equal parts. Find the angles of the triangle.

Logic

L.1. There are three parrots, A, B, and C. One of them tells only the truth, another always lies, and the third is a sly one and sometimes tells the truth and sometimes lies. When asked: "What type is parrot B?" the three parrots replied as follows. A: "B is a liar." B: "I'm the sly one." C: "B is an absolutely truthful parrot." Which of the three parrots is a liar, and which is the sly one?

L.2. Three friends with the last names Brown, Black, and Red got together one day. One of them, who had black hair, pointed out: "One of us has brown hair, one has black hair, and one has red hair, but none of our last names corresponds to his hair color." "You are right" said Brown. What color hair does each of the friends have?

L.3. Four girls — Ann, Beth, Sarah, and Hope — are standing in a circle and talking. The girl in a green dress, who is neither Ann nor Beth, is standing between the girl in a blue dress and Hope. The girl in a white dress is standing between the girl in pink and Beth. What color dress each girl is wearing?

It is often convenient to place the information in a table like the one below. In the table there are two ×'s to indicate that the girl in a green dress is neither Ann nor Beth. All that's left to do is to use the rest of the conditions to rule out other combinations and solve the problem.

	White	Blue	Green	Pink
Ann			×	
Beth			×	
Sarah				
Hope				

L.4. Three clowns, Bimbo, Bingo, and Bongo, came into the circus ring wearing red, blue, and green shirts. Their shoes were also red, blue, and green. Bingo's shirt and shoes were of the same color. Bongo was wearing nothing red. Bimbo's shoes were green and his shirt was not. What colors were Bongo's and Bingo's shirts and shoes?

L.5. A bottle, a glass, a carafe, and a jar contain milk, lemonade, Coke, and water. Neither water nor milk are in the bottle. A vessel containing lemonade stands between the carafe and the vessel which contains Coke. The jar contains neither lemonade nor water. The glass stands beside the jar and the vessel that contains milk. What is contained in each vessel?

L.6. Each one of four gnomes, Bim, Tim, Kim, and Sim, either always tells the truth or always lies. We have overheard the following exchange.

Bim said to Tim: "You're a liar!" Tim said to Kim: "You're a liar yourself!" Kim said to Sim: "They're both liars. And you, too."

Which gnomes always tell the truth?

L.7. A cabinet consists of 20 ministers. At least one of them is honest. In every pair of ministers, at least one is corrupt. How many ministers are honest?

L.8. Susan and Sam went to a stationery store to buy a sketchbook each. Sam was seven dollars short, and Susan one dollar short. They decided to buy just one sketchbook between them and split the pages. After they pooled their money together, they still didn't have enough even for one sketchbook. How much did a sketchbook cost? (The price is known to be a whole number of dollars.)

L.9. Everyone in Truthville always tells the truth; everyone in Liartown always lies, and everyone in Fickle City answers questions alternately with the truth or a lie. How can you pose exactly four questions, each having a yes or no answer, to a person you on a street and decide what city you are in?

L.10. Three runners took part in a race. Before the race, four spectators made the following predictions:

Smith will be the winner.
Okuda will finish before Chavez.
Chavez will finish right after Smith.
Okuda won't win.

After the race was over, it turned out that an even number of predictions were true. In what order did the runners finish?

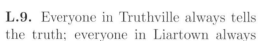

L.11. Four cards lie on a table as shown above. Each has a letter on one side and a number on the other. Your classmate tells you that if there is an even number on one side of a card, there is a vowel on the other side. Which card(s) must be turned

LOGIC

over in order to find out whether this is true? (Б is the Russian letter that sounds like B, so it's a consonant.)

L.12. Imagine that you are on an island where all the natives are either knights, who always tell the truth, or knaves, who always lie. A person says "I am a liar". Is he a native?

L.13. Imagine that you are on an island where all the inhabitants are either knights, who always tell the truth, or knaves, who always lie. There is a crowd of people on a plaza, and each person in the crowd declares "You are all liars". How many knights are there?

L.14. Imagine that you are on an island where all the inhabitants are either knights, who always tell the truth, or knaves, who always lie. Seven people sit around a table, and each of them says "One of my two neighbors is a liar and the other is a knight". Who is sitting at the table?

L.15. Imagine that you are on an island where all the inhabitants are either knights who always tell the truth, or knaves, who always lie. Moreover, all the knights live in one town and all the knaves live in another. If you are on a road and meet an inhabitant, how can you find out if the road goes to the knights' town or the knaves' town?

L.16. The figure shows Amanda, Betty, Colette, and Deborah, from left to right. Which two girls should be asked to turn around if you want to check whether this statement is true: If a girl is not wearing glasses, there is a bow in her hair.

L.17. Three boxes contain some balls. There are two white balls in one box, two black balls in another box, while the third box contains one white and one black ball. The boxes are labeled. On one box the label says WW, on another it says BB, and the label on the third box is BW. None of the labels describes the contents of the box correctly. How is it possible to decide which box contains what by taking exactly one ball out of a box?

L.18. Gertrude likes pets. All but two of her pets are parrots. All but two are cats. All but two are dogs. All the rest are crickets. How many pet crickets does Gertrude have?

L.19. In a certain country there are three towns: Alitheia, Mentira, and Twijfel. In Alitheia everyone always tell the truth; inhabitants of Mentira always lie; and people living in Twijfel alternately tell the truth and lie. Once a fire broke out in one of these towns, and the fire department was called. There ensued the following dialog:

"There is a fire here!"
"Where is it?"
"In Twijfel."

Where should the firefighters go?

L.20. A fashion shop has just received a new line of dresses. The staff wants to showcase the three models and three colors from that line, all of which are represented in the shipment. Are they always able to choose three dresses from the shipment to represent all models and all colors?

L.21. Pete's cat always sneezes 24 hours prior to a rainstorm. Today his cat sneezed. "It will rain tomorrow," thought Pete. Is he right?

L.22. A box contains pencils of different lengths and different colors. Prove that there are two pencils that differ both in length and in color.

Combinatorics

C.1. Each of three bookcases contains 100 books. All of these 300 books are different. How many ways can two books be chosen if they must come from different bookcases?

C.2. How many diagonals are there in a convex polygon with 100 sides?

C.3. How many 5-digit numbers with at least two identical digits are there?

C.4. In how many ways can a black and a white square on a chessboard be picked so that the chosen squares do not lie in the same column?

C.5. Suppose that we have fabric in five different colors. How many different striped flags can be made if each flag must consist of three horizontal stripes of different colors?

C.6. Suppose that we have fabric in five different colors. How many different striped flags can be made if each flag must consist of three horizontal stripes of different colors with one of the stripes red?

C.7. In how many ways can a striped flag can be designed if it must consist of 11 stripes each colored either red, blue, or white, and any two adjacent stripes must be different colors?

C.8. Ten ladies and ten gentlemen attend a ball. In how many ways can they split into pairs consisting of a lady and a gentleman?

C.9. How many words can be made out of five A's and at most two B's? In this problem a word is any sequence of letters.

INTENSIVE CARE

C.10. In England it is customary to give a child from one to three names. How many ways a child can be named if there are three hundred English names?

C.11. In how many ways can six urgent letters be sent by six couriers if each letter can be given to any of the couriers?

C.12. In how many ways can five men and five women be seated around a table so that no two people of the same gender sit next to each other?

C.13. Pete has five math books, and Jorge has seven. In how many ways can they swap two books? Swapping two books entails each giving the other two books.

C.14. You have three flags and can raise at least two of them up a flagpole as a signal. How many different signals can you send if the order of flags in each signal matters?

C.15. An ice cream van is carrying chocolate, strawberry, and vanilla ice cream. How many different three-scoop cones can you buy?

C.16. How many seven-digit numbers do not contain the digit 2?

C.17. What is the number of six-digit numbers that have at least one even digit?

C.18. Two rooks on a chessboard threaten one another if they are in the same row or in the same column. In how many ways can eight rooks be placed on a chessboard so that they do not threaten one another?

Puzzles

It's not hard to choose some relatively easy mathematical puzzles that should be solved as normal mathematical problems. However, it is best to stock up on real puzzles made of wood or metal, where the goal is, for example, to build a pyramid out of given pieces or to untangle some rings.

Intensive Care

This is a room where students were sent if they couldn't solve a problem given to them in the main room. Here the unsolved problem was replaced with a very easy one. Problems of the "Kangaroo" contest would be suitable for this room.

Two and Two Is More Than Four: A Story

This text is a composite of several conversations among students, overheard by the author during actual circle sessions. It is meant to give both students and prospective circle leaders a feel for the atmosphere of lively discussion, collaboration, and teasing that characterizes many circles, and which can make teaching such a circle a worthwhile experience. It is immensely rewarding to watch students interact and build on each other's ideas.

Several of the problems were given in the text. Problem 1 of the story is 15.5 in the text, Problem 2 is 18.3, Problem 3 is 19.7, Problem 5 is 0.6, and Problem 6 is 16.2.

I. Sail Me Down the River[1]

Andy, Dan, Beth, and Ted came to their math circle after school. There was still time before the start, and the teacher wasn't there, but somebody had already written some problems on the blackboard. This was the first:

Problem 1. *Two boys simultaneously jumped off a raft floating on a river and started swimming in different directions: one downstream, the other upstream. In five minutes they turned around and soon were back on the raft. Assuming each kept a steady pace, which one got back first?*

"I wonder, is this problem for us?" asked Beth.

"Wouldn't it be so much more fun to go to the river than to do these problems?" replied Ted. "And this problem is wrong anyway."

"Why?" asked Beth.

"They also need to state that the boys' speeds were greater than the speed of the river's current," said Ted, "because otherwise the one who started upstream would not be able to swim away from the raft at all."

"Well, I think you're wrong!" Dan objected. "Even if his speed is the same as the river's current, he will swim away from the raft."

"Oh, come on!" protested Ted. "He will then flounder at the same spot! Once this happened to me. I kept swimming as fast as I could, and there

---

[1] A quote from the song "Down by the Sea", by the band Men at Work.

was a big pine tree opposite to me. I swam and swam, but still remained where the tree was. I hardly made it to the shore then. The river was too fast."

"You may have stayed at that spot, but the raft won't wait for you," said Dan, "it isn't on the shore like your pine tree, the river will pull it away."

"Now I'm totally confused," Ted said, "so, even if I'm slower than the river, I'll still be able to swim away from the raft, will I?"

"Sure! Because the raft isn't resisting the current, and you, though doing so weakly, are paddling against it."

"Well, then how do we solve the problem?"

Everyone was silent for a while with each trying to think it through, but no one came up with a solution.

"Oh, listen, I have an idea," Andy said suddenly. "Suppose it all happens on a lake. It is clear that both will return simultaneously. The raft stays still, everyone is swimming for five minutes away from it, and then returns for just as long."

"This isn't a lake though, it's a river," objected Dan. "It flows!"

"So what! The river does the same thing to everything in it; the river moves it downstream with the speed of the current. It is as if the current isn't there at all."

"This sounds right. Could it be so simple?"

"Of course. And in the fog, when you can't see the riverbank, you won't distinguish between a river and a lake at all."

"I've got it, I've got it!" Everyone looked at Ted, who, in his excitement, even started flapping his arms. "I can explain it in a simpler way. My father is a cameraman. He was filming a thriller, and in one scene he had to follow a boat in the river with a camera along the riverbank. So if he had always remained opposite to the raft while filming, what would we see on the screen?"

"Not much." Dan shrugged his shoulders. "In the middle of the screen there would be a raft, and the opposite shore with the trees would be moving, like in a train's window."

"At the moment, we don't care about the bank." Ted said sternly, "What's important is that there will be a still raft in still water. Two boys jumped from the raft in different directions. Each swam his distance for five minutes, and then he would swim back in the same five minutes."

"Yes, this is great," said Andy. "I think this is called changing the frame of reference to one that is based on the river. My brother studied this in his physics class."

"Don't be a smart aleck. But about the train, that was clever! Why didn't it occur to us?" Ted was glowing with pride. "The river moves everything forward like a train or an airplane. If we started running in opposite directions in the aisle of a flying airplane, and turned around in a couple of seconds, we would meet where we started."

1. Sail Me Down the River

"Yeah," agreed Dan. "And if you jump in an airplane, then you'll land in the same place, and, even though the airplane is moving at a huge speed, it won't fly out from under you."

"This is because," continued Andy, "the airplane's speed relative to the ground may be high, but you are moving at the same speed, so for you, the airplane doesn't move."

"Actually, when I was flying in a plane," said Ted, "during the takeoff I was being squished into the chair, and, if at that moment I was in the aisle and tried to jump, then I would have found myself in the tail of the plane."

"That is because, at that moment, the airplane was gaining speed and moving with acceleration," answered Dan. "That is a completely different thing. No wonder that walking is not allowed during takeoff and landing."

"Now I understand why cowboys in the movies aren't afraid to run across the tops of the trains and jump from car to car;" continued Ted, "it's the same as running on a stationary one."

"Well, not quite," said Andy, "in a train you have counter air pressing on you. When you jump, the air immediately slows you down — the air isn't moving with the train. If the train is moving very fast, the wind may blow you off the roof. In the time of cowboys the trains were very slow, even an average horse could outrun a train."

"While you were arguing about your cowboys, I solved the problem algebraically, with the help of equations," said Beth, waving a piece of paper. "Do you want me to show you?"

"Everything is clear anyway. Well, show us, so you haven't worked for nothing." The kids gathered around Beth's piece of paper.

While pointing to what she wrote, Beth continued. "It is short. In meters per minute, let v be the speed of the river, and x the speed of the first boy. He swims downstream for five minutes with the speed $x+v$ with respect to the bank. He will be $5(x+v) - 5v = 5x$ meters from the raft. The next five minutes he will be moving at $x-v$ and will swim $5(x-v) = 5x - 5v$ meters. So at this point he will meet the raft, which would have floated toward him $5v$ meters during these five minutes. The same can be checked for the second one."

"We've got the same thing, only without any x and v."

"Let's also solve the second problem."

Problem 2. *Two towns are located on a river, 10 miles apart along the water. Will it take a ship longer to go from one town to another and back, or to cover 20 miles on a lake?*

"Well, this is an even simpler problem," Ted declared at once. "When the boat sails downstream, the river helps the boat, and when it goes upstream, the river hinders it and eats up all the benefits gained by the help. Thus sailing downriver and back again upriver is the same as sailing on the lake."

"Something's wrong here," said Andy with a hint of doubt in his voice. "Imagine that the speed of the boat is equal to the speed of the river. Then the boat will never get back to the upriver town — it won't be able to fight the current. And on the lake, no problem! So it looks like it is always faster on the lake."

"But what's my error?"

"I think I've got it. You say that when the boat sails down the river, the current helps it, but what does this mean?"

"Well, isn't it clear?"

"'Helps' means that every second the current moves the boat some meters ahead; 'hinders' means that every second it moves the boat the same number of meters back compared to on the lake."

"And I said the same thing, only I said it shorter and clearer."

"Yes, but you forgot the main thing. The boat sails the same distance up and down the river, but it moves down faster than up, so the river helps the boat less than it hinders! Thus, there is less benefit than resistance."

"Oh, I am a fool! Yeah, great! So it will be faster on the lake."

"I've solved the problem algebraically," said Beth.

"Listen, you, 'guru-algebraist'," Ted needled her, "what is there to write if everything is already clear?"

"You got lost, so I wrote the equations. How could I know that you'd get it so quickly?"

"Okay, show us. Do we have to check it?"

Here was what Beth had written down:

v speed of the boat

u speed of the river (in km/h)

10 km downstream: $10/(v+u)$ hours

10 km upstream: $10/(v-u)$ hours

Total: $10/(v+u) + 10/(v-u) = 20v/(v^2 - u^2)$ hours

20 km on the lake: $20/v$ hours

$$\frac{20v}{v^2 - u^2} > \frac{20v}{v^2} = \frac{20}{v}$$

"I have trouble with algebraic computations," Ted said. "I'd rather do without them. Okay, do we have anything we have not solved yet?"

There was one more problem on the board.

Problem 3. *A boat goes downriver from town B to town B in three days, and goes upriver from B to A in five. How long will it take a raft to float from A to B?*

"Oh, now we're *really* going to need an equation." Ted became gloomy. "Here it says that the boat sails down the river for three days and up the river for five. And what does that mean?"

I. Sail Me Down the River

"Just that, relative to the ground, the ratio of their speeds is 5 to 3," said Andy.

"Whose speeds?" Ted sounded confused.

"Well, of the boat that sails downstream and one that sails upstream," Andy clarified. "One can even say that we sail two identical boats: one downstream, and another upstream. Then, if the first one sails 5 km on the river, the second will sail 3 km in the same time."

"Wow! I had a great idea!" Dan jumped in.

"What?"

"Send two boats simultaneously in different directions, but both starting from A. We also need to send the raft downstream."

"Why?" Ted was surprised.

"Because the raft will be midway between the boats all the time, if we measure distance along the river! Remember the problem with the boys jumping off the raft."

"Wow, cool! The boats are like those two boys! And since the boats are identical, they are sailing away from the raft with the same speed. But how does this help us?"

"Here's how. If the first boat sailed $5x$ km away from A, then the second one sailed $3x$ km. Do you agree? The raft is half way between them all the time. And where will this midpoint be?"

"The distance between the boats is $8x$ km; half of that is $4x$ km. So the raft will be x km downstream from A."

"And here we are. We want to know when the raft will reach B, so x should be taken equal to the distance between A and B. The first boat will sail 5 times the distance from A to B in this time. It takes 3 days to sail from A to B, and 5 times this time is 15 days. That means that 15 days is the answer."

"I wrote an equation; it happens to be easy, too." Beth said.

"You can't live without your equations," Ted said in frustration.

"Just look. It is only three lines. Let's say that the distance between A and B is 1."

"I didn't get it. One what?"

"Who cares! There is some distance between them, and we can say that's the unit of measurement. Remember that cartoon we once saw where the length of a snake was measured in parrots? We will measure lengths in distances between A and B. It is just as good as in kilometers. It is always done for convenience so as not to introduce an extra variable."

"Okay, what's next?"

v speed of boat

u speed of stream

$3(u+v) = 1$ and $5(v-u) = 1$

$u+v = 1/3$ and $u-v = 1/5$

subtract second equation from the first

$2u = 1/3 - 1/5 = 2/15$, so $u = 1/15$

"So the raft will float from A to B in 15 days."

"And our answer is the same."

"And I've invented a new problem myself! Hooray!" Andy ran to the board and grabbed a piece of chalk.

"Really?"

"Yes. I liked our first solution to the boat problem so much that I made up a problem of my own. Solve it!" And Andy wrote his problem on the board.

Problem 4. *From point A a raft and a boat sailed simultaneously downstream on a straight river, and at the same time another boat of the same type sailed from point C towards them. Prove that at the moment when the first boat reaches point C, the raft will be exactly midway between point A and the second boat.*

"It's almost the same problem," shouted Dan.

The children quickly solved Andy's problem, and were very fond of it. Can you solve it too?

II. Tanks and Escalators

At this moment Sergey, the math circle teacher, entered the room. The students started explaining their solutions to him, while vying with one another for his attention. Then they gave him the problem Andy posed, and carefully checked his solution.

"Well done," the teacher said. "It can be said that you managed to run a math circle without me, and you even came up with a great problem. But I also have something for you."

Problem 5. *The diagram shows the rolling track of a tank, seen from the side. The bottom is in contact with the ground. If the tank moves forward 10 cm, how many centimeters does the point marked A move?*

"It's a problem for first graders, isn't it?" wondered Ted. "If the tank traveled 10 centimeters, then point A also traveled 10 centimeters."

II. Tanks and Escalators

"If that were so, then point A would have been opposite the middle of the tank all the time, wouldn't it?" Dan asked.

"So what?"

"But, you see, the point is on the track. It would turn out that the middle of the tank and the middle of the track have been aligned with each other all the time. It means that the track did not move. Then how could the tank have moved?"

"That's true, the tracks have to spin," Ted grew thoughtful. "It's the same thing with tanks in the movies. It seems the tracks are spinning too fast, even faster than the tanks move forward. So point A is moving faster than the tank itself."

"Hey, let's ask your father, the cameraman, to help us. He'll ride opposite the tank with a camera and film it."

"Okay," Ted perked up. "Then we will see the unmoving tank with just the wheels rotating and tracks moving. For the tank to move 10 centimeters, 10 new centimeters of the track should touch the ground, so the track has to spin forward 10 centimeters. That means that point A will also move forward on the screen 10 centimeters. Once again, we get the same darn 10 centimeters."

"But the camera also moved 10 cm! So A moved a total of $10 + 10 = 20$ centimeters."

"Amazing! It's hard to believe."

"Nevertheless, this is the answer, unless it's a toy tank and point A moved around the front wheel," confirmed Sergey. "Now I will try to surprise you once again. Here's a new problem."

Problem 6. *Pete and Jack are riding down an escalator. Halfway down the escalator, Jack grabbed Pete's hat and threw it onto the up escalator. Pete ran up the down escalator in order to run down the up escalator after his hat. Jack ran down the down escalator in order to run up the up escalator after the hat. Who will be first to the hat? Assume the speeds of the boys relative to the escalator are equal and do not depend on the direction of motion.*

"Well, well," Ted started talking to himself. "The escalator first helps Jack, and then, when he changes to the up escalator, oh, it helps again. On the other hand, the escalator hinders Pete. So, Jack will get the hat first."

"Wait a minute, but the hat is going up, towards Pete, and at the same time it moves away from Jack," noted Beth.

"Then Pete will be the first. No, wait, I'm confused," Ted said, "Maybe they will get to the hat simultaneously?"

"If the escalators were not moving, then of course, they would arrive at the same time; each must cover two halves of an escalator."

"Probably here, just like in the river problem, we can assume that the escalators are not moving."

"Exactly!" Dan joined the conversation. "The escalator hinders as well as helps; it is as if it moved both Jack and the hat forward while moving Pete and the hat backward."

"Yes, I get it," Ted agreed. "So the distance between the hat and either one of the boys is getting smaller at the same rate as if the escalators were not moving."

"Well done," said the teacher. "Let me put your arguments a little differently."

The teacher came to the board and drew two semi-circles with arrows.

"These are the escalators," he said, "and the arrows show the way they're moving."

"*These* are escalators? That's funny," The students laughed.

"You'll soon understand. I drew them curved like this because when one of the boys runs off his escalator he moves onto the other. Let us connect these half-circle escalators into a circle. That means that one escalator turns into the other, and the boys are running on a moving circle."

"Hooray! Now it is clear. First Pete with Jack and the hat will be at the opposite points of the circle. And then they will run to the hat in opposite directions with equal speeds. It is clear that they will get to the hat simultaneously. And it does not matter whether the circle spins or not. Give us the next problem!"

"Wait, we still have not finished this one."

"How is that?" Ted was very surprised. "We've proved that they will come at the same time, what else is needed?"

"This answer is not quite correct."

"What?! But we have proved the answer — in two different ways, in fact. Are you kidding?"

"No. We've missed something."

"True, there's also the distance between the escalators," mused Dan, "When they run the distance it's as if they temporarily jump off the circle while the hat continues towards Pete. Thus he'll be the first."

"A very thoughtful observation. But even if the boys cover the distance between the escalators instantaneously, there is still a possibility of one of the boys getting to the hat before the other one."

"Ah, yes!" surmised Ted. "If Peter's speed is less than or equal to the speed of the escalator, he will not reach the top of the escalator, and the hat will not jump to him!"

"Yes, but that's not all; even if the boy's speed is greater than the speed of the escalator, a different answer is still possible."

Oh, I got it!" Andy shouted. "Ted just gave us a hint. If the boys run too slowly, the hat can reach the top of the escalator."

"And then who will get it first?"

"The hat will stop coming to Pete and no longer move away from Jack. It means, Jack will get it first."

"Yes, now all the possibilities have been covered. Good job! That's all for today," said Sergey.

Addendum:
The San Jose Experience

by Tatiana Shubin

In November 2006, a group of nineteen U.S. math circle enthusiasts went to Russia on a trip to observe Russian math circles and to learn from their long, rich, and very fruitful experience. The trip was organized by MSRI and cosponsored by NSF. We spent a day in St. Petersburg and three days in Moscow visiting various specialized math schools, talking with students and instructors, sitting at the back of classrooms through their regular math classes, and participating at math circle meetings—sometimes just watching, and, at times, even actively participating as additional instructors at their math circle meetings. Our participation was made possible by the unusual structure of the Russian meetings. The structure was very different from what we had been used to in our own circles back home. It was not a lecture, nor was an instructor in the front of a classroom directing activities. Instead, students were given "scripts"—lists of problems to work on individually, and when a kid wanted to talk about his or her solution, an instructor would sit next to them and listen. Eventually, the classroom was abuzz with conversation—every student had a mathematician to talk to and to help in wrestling with the problems, or better yet, finding interesting ramifications or elucidating some deep implications.

Needless to say, we were profoundly impressed. One conversation in particular made me think of a way to try and import this kind of operation. Nikolai Konstantinov, one of the most influential people in the math circles of Russia, once said, "The greatest reward for a good student is not a good grade. It's the willingness of his teacher to listen to him."

As you can imagine, to conduct a circle this way is very labor-intensive: the student–instructor ratio should be somewhere near two to one. It's impossible for us to hold such meetings every week. But we found a compromise—twice a year, once in the fall semester, and once in the spring, we hold what we call a Russian-style math circle. Below is the description from our website (sanjosemathcircle.org).

"Russian-style math circles are different from our usual meetings. In a Russian-style session, it is the students who talk! Here's how it works: We will post a set of problems online, usually about a week before the Russian session takes place, and you, the students, try to solve as many problems as you can at home. Then, at the session you can present your partial or complete solutions, ask questions, etc. You'll be talking to math teachers who can follow your thinking and make suggestions about how to improve your presentation. You will be getting as much help as you want, as well as having the experience of explaining your ideas to others. But relax: if you can't solve some of the problems, we are there to help. If you do solve all of the problems, we'll have more of them waiting for you at the circle meeting! The idea is to help you learn to present your ideas and to learn some mathematics, too. Nobody is expected to solve all of the problems; just do your best, come to the meeting, and we'll be there to help!"

These meetings are very popular with the students who appreciate both the challenge and the chance to talk. We believe that it is a valuable experience for the kids, and it's fun for the instructors, too. Over the years, we have had a great many helpers so it is impossible to name all of them here. But at least some of them should be mentioned. Tom Davis, a co-founder and co-director of the San Jose Math Circle, has been co-hosting the event starting with the very first Russian-style meeting in December of 2006. Alon Amit, Lucian Sega, and Paul Zeitz helped to compile sets of problems, and also helped at the meetings. Paul Zeitz and Matthias Beck brought their students from University of San Francisco and San Francisco State University, respectively, to help run some Russian-style sessions. Finally, we were very lucky to have Vladimir I. Arnold come to our Russian-style meeting in March 2007. In the photograph, he is working with one of our youngest circlers.

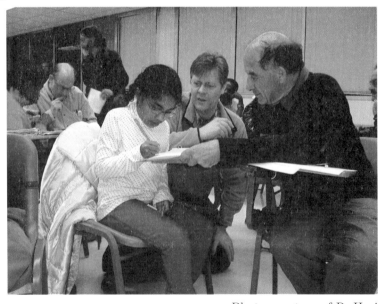

Photo courtesy of R. Hari

Addendum: The San Jose Experience

Over the years, the structure of these meetings has been evolving. Now we give students a few problems to work on ahead of time — it's their "homework" — but when they arrive at the meeting they are given a set of new problems and they work on them for about 40 minutes on the spot. We are still experimenting on comparative length and difficulty of problems, as you can see from the sample sets given on the next few pages. Our meetings are two hours long; the majority of our students are seventh and eighth graders. These sets of problems worked pretty well for our circle.

The problems in these sets are not original — we borrow them from lots of sources. Selecting good problems at a right level and enough topics to make it exciting for students of different ages, with varying abilities and backgrounds, is a challenging problem in itself. We hope that these sets will be useful for people who would be interested in conducting their math circles in the Russian style.

Problem Set SJ1

Homework Problems

Instructions for students: Work through as many problems as you can. Even if you can't solve a problem, try to learn as much as you can about it.

Problem SJ1.1. To "tile the plane" means to cover it with geometric figures without any overlap or gaps. Given a convex quadrilateral, is it always possible to tile the plane by quadrilaterals that are all congruent to the given one? Does there exist a convex pentagon with pairwise nonparallel sides such that it is possible to tile the plane by its copies?

Problem SJ1.2. On a number line color red all the points that can be expressed in the form $81x + 100y$, where x and y are positive integers, and color all other integers blue. (So 262 is red, but 100 is blue.) Prove that there exists a point P on the number line with the property that any two integer points which are symmetric with respect to P are of different colors.

Problem SJ1.3. If the expression with k pairs of parentheses

$$(\ldots(((x-2)^2 - 2)^2 - 2)^2 - \cdots - 2)^2$$

is multiplied out, what is the coefficient of x^2?

Problem SJ1.4. Three players Al, Bob, and Cy play the following game. On each of three cards an integer is written. These three integers p, q, and r satisfy the inequalities $0 < p < q < r$. The cards are shuffled and one is dealt to each player. Each one then takes the number of counters indicated by the card he holds. Then the cards are shuffled again, while the players keep their counters. This process—shuffling, dealing, and taking counters—takes place at least twice. After the last round, Al has 20 counters, Bob has 10, and Cy has 9. On the last round, Bob took r counters. Who got q counters on the first round?

Problem SJ1.5. A triomino is a 1×3 rectangle consisting of unit squares. Suppose that an 8×8 chessboard made of unit squares has 21 triominos placed on it in such a way that each triomino covers exactly three squares of the board. Naturally, one of the squares is left uncovered. Which of the chessboard squares could it possibly be?

Problem SJ1.6. Two regular pentagons are situated above one another in space. In addition to the ten edges already present, every edge joining a vertex of the upper pentagon to a vertex of the lower pentagon is also drawn, making a total of 35 edges. Each edge is then colored either red or green in such a way that every triangle formed by the edges has at least one edge of each color. Show that all ten edges of the two pentagons must be the same color. Would this conclusion still be true if the top and bottom shapes were n-gons, where $n \geq 3$ is a positive integer?

Problem SJ1.7. When 4444^{4444} is written in decimal notation, the sum of its digits is A. Let B be the sum of the digits of A. Find the sum of the digits of B.

Problem SJ1.8. Once upon a time there was a long and fierce war between two clans, the Blue Iris and the Purple Iris. At a time of truce, knights of both clans sat together around a huge round table. It turned out that the number of knights who had a friend sitting next to them on their right was the same as the number of knights with an enemy on their right. Could the total number of knights around the table be 1386?

In-Class Problems

Problem SJ1.9. Find three natural numbers which form an arithmetic sequence and whose product is a perfect square.

Problem SJ1.10. Is the number $4^{545} + 545^4$ prime or composite? Justify your answer.

Problem SJ1.11. A game board is a long 1×100 rectangle made of one hundred 1×1 squares. A single marker is placed in the leftmost square. Two people take turns moving the marker; it can be moved 1, 10, or 11 squares to the right. The player who is unable to move the marker loses the game. Which of the two players has a winning strategy? How should he play in order to win?

Problem SJ1.12. Suppose we are given line segments of lengths 1, 2, 3, ..., 99. If we have to use all 99 segments, is it possible to construct a square? How about a rectangle?

Problem SJ1.13. Suppose that $ABCD$ is a square. P is a point on side AB such that $\frac{AP}{PB} = \frac{3}{2}$; Q is a point on diagonal AC such that $\frac{AQ}{QC} = 4$. Find the angles of triangle PDQ.

Solutions to Problem Set SJ1

Problem SJ1.1. Let the angles of the quadrilateral be a, b, c, and d. Since $a+b+c+d = 2\pi$, we can place four copies of the quadrilateral with these four angles meeting at a point, so that the edges of adjacent figures match and the angles of the four figures starting at the common vertex and going counterclockwise around each figure are c, d, a, b; b, c, d, a; a, b, c, d; d, a, b, c. Clearly, this configuration can be repeated at every vertex producing the desired tiling.

The answer to the second question is yes. Take a regular hexagon and split it into three congruent pentagons by three line segments from the center to the midpoint of every other side of the hexagon.

Problem SJ1.2. Let $S = \{81 \cdot 1, 81 \cdot 2, 81 \cdot 3, \ldots, 81 \cdot 100\}$. Clearly, elements of S are incongruent modulo 100 and since $|S| = 100$, S is a complete residue system modulo 100, i.e., for every $0 \leq n \leq 99$ there is exactly one $s \in S$ such that $s \equiv n \pmod{100}$. Therefore, if $N \geq 8100$, then $N = 81x + 100y$, where $1 \leq x \leq 100$, and $y \geq 1$. On the other hand, if $8100 = 81x + 100y$, then $81|y$, and $100|x$, so that $8100 = 8100a + 8100b$ for some $a \geq 1$, $b \geq 1$, a contradiction. Thus 8100 is the largest blue number. Obviously, 181 is the smallest red number. Let $P = \frac{181+8100}{2} = 4140.5$. If r is a red number, then $r = 181 + 81m + 100n$, where m and n are nonnegative numbers. Let t be symmetric to r with respect to P. Then $P = \frac{r+t}{2}$, $t = 2P - r = 181 + 8100 - 181 - 81m - 100n$ yielding $8100 = t + 81m + 100n$. Hence if t is red, 8100 would be red, too; a contradiction. Therefore t is blue. Thus P is the desired point.

(Note: the proof is incomplete — one should also prove that if r is blue, t is red.)

Problem SJ1.3. Let's denote the polynomial in the problem $P_k(x)$, and write $P_k(x) = a_{n,k}x^n + a_{n-1,k}x^n + \cdots + a_{2,k}x^2 + a_{1,k}x + a_{0,k} = Q_k(x)x^3 + a_{2,k}x^2 + a_{1,k}x + a_{0,k}$. Then $P_1(x) = (x-2)^2 = x^2 - 4x + 4$, and so $a_{0,1} = 4$, $a_{1,1} = -4$, $a_{2,1} = 1$, $Q_1(x) = 0$. Let us derive a recursion for $a_{2,k}$. Write

$$P_k(x) = (P_{k-1}(x) - 2)^2 = (Q_{k-1}(x)x^3 + a_{2,k-1}x^2 + a_{1,k-1}x + a_{0,k-1} - 2)^2$$
$$= Q_k(x)x^3 + 2a_{2,k-1}(a_{0,k-1} - 2)x^2 + (a_{1,k-1})^2 x^2$$
$$+ 2a_{1,k-1}(a_{0,k-1} - 2)x + (a_{0,k-1} - 2)^2.$$

Hence
$$a_{0,k} = (a_{0,k-1} - 2)^2 = (a_{0,k-2} - 2)^2 = \cdots = 4,$$
$$a_{1,k} = 2a_{1,k-1}(a_{0,k-1} - 2) = 4a_{1,k-1} = \cdots = 4^{k-1}(a_{1,1}) = 4^{k-1}(-4) = -4^k,$$
$$a_{2,k} = 2a_{2,k-1}(a_{0,k-1} - 2) + (a_{1,k-1})^2 = 4a_{2,k-1} + 4^{2(k-1)}.$$

Repeated application of the last relation gives
$$\begin{aligned} a_{2,k} &= 4(a_{2,k-2} + 4^{2(k-2)}) + 4^{2(k-1)} \\ &= 4^2 a_{2,k-2} + 4^{2k-3} + 4^{2k-2} = \cdots \\ &= 4^{k-1} a_{2,1} + 4^k + \cdots + 4^{2k-3} + 4^{2k-2} \\ &= 4^{k-1} + 4^k + \cdots + 4^{2k-2} \\ &= \frac{4^{k-1}(4^k - 1)}{3}. \end{aligned}$$

Problem SJ1.4. Let $N \geq 2$ be the number of rounds played. Then $(p + q + r)N = 39$. Since $0 < p < q < r$, we conclude that $p + q + r \geq 1 + 2 + 3 = 6$, and since $N = 39 = 3 \cdot 13$, hence $N = 3$, $p + q + r = 13$. Denoting by a_i, b_i, c_i the number of counters taken by Al, Bob, and Cy in the i-th round, we tabulate the information we have so far, keeping in mind that each entry is one of the numbers p, q, and r, and that the entries in each row are distinct.

Round	Al	Bob	Cy	Total
1	a_1	b_1	c_1	13
2	a_2	b_2	c_2	13
3	a_3	r	c_3	13
Total	20	10	9	39

Since $a_3 \neq r$, the sum of the numbers in the first column is at most $2r + q$. Hence $2r + q \geq 20$. Substituting $q = 13 - p - r$ in this inequality, we obtain
$$r - p \geq 7. \tag{1}$$
Since $p \geq 1$, this implies that $r \geq 8$. On the other hand, the sum of the numbers in the second column is at least $r + 2p$. Therefore,
$$r + 2p \leq 10. \tag{2}$$
This inequality together with $p \geq 1$ implies that $r \leq 8$. Thus $r = 8$, and we can conclude from either equation (1) or (2) that $p = 1$. Finally, $q = 13 - p - r = 4$. Since the equality holds in equations (1) and (2), our bounds for the column sums are sharp. Hence the solution is this:

Round	Al	Bob	Cy
1	8	1	4
2	8	1	4
3	4	8	1

In particular, it was Cy who got $q = 4$ counters in the first round.

Problem SJ1.5. In what follows, square nm denotes the square in row n and column m of the board. Rows are counted from top down; columns are counted from left to right. So, as an example, square 11 is the upper left square of the board.

Let's assign one of the letters A, B, C to each square, as follows. Label square 11 by A; label squares 21 and 12 by B; label squares 31, 22, and 31 by C; continue by labeling each rising diagonal of the chessboard A, B, C in rotation. After that, there are 21 A's, 22 B's, and 21 C's. Each triomino necessarily covers squares containing all three labels; thus the only squares that might be left uncovered are those labelled B.

Next we relabel all the squares: instead of starting at square 11 we start at 81 and label it A; squares 71 and 82 get B; and we proceed labeling as before, but this time using all falling diagonals. The argument above applies again, and the only squares that are labelled by B's both times are squares 33, 36, 63, and 66.

Now it's easy to construct tilings that leave each of these squares uncovered. For example, place triominos vertically on squares 11, 21, 31; 12, 22, 32; 37, 47, 57; 38, 48, 58; 67, 77, 87; 68, 78, 88, and place triominos horizontally elsewhere. The rest of the squares—36, 63, and 66—can be left uncovered by the same scheme, altered by a suitable symmetry.

Problem SJ1.6. Let's denote the vertices of the top and bottom pentagons by $A_1A_2A_3A_4A_5$ and $B_1B_2B_3B_4B_5$, respectively. Let us now show that all edges A_iA_j are of the same color. Suppose not, and, say, A_1A_2 is red while A_2A_3 is green. At least three of the five segments A_2B_1, A_2B_2, A_2B_3, A_2B_4, A_2B_5 have the same color. Let's assume without loss of generality that they are red, and label them A_2B_i, A_2B_j, A_2B_k. Then at least one of the segments B_iB_j, B_jB_k, B_kB_i is an edge of the base; call it B_rB_s. If B_rB_s were red, we would have a red triangle $A_2B_rB_s$. Therefore B_rB_s is green. Now segments A_1B_r and A_1B_s must also be green, for otherwise at least one of the triangles $A_1A_2B_r$ or $A_1A_2B_s$ would be a red triangle. But then $A_1B_rB_s$ is a green triangle. This contradiction implies that A_1A_2 and A_2A_3 have the same color; likewise, all edges of the top pentagon are of the same color. Similarly, all the edges of the bottom pentagon also have the same color.

Now suppose that the top edges are all red and the bottom edges are all green. If three green edges join A_1 to the bottom, two of them must terminate at the adjacent vertices B_r, B_s of the base. Then $A_1B_rB_s$ is a green triangle. Thus at least three red edges join A_1 to the bottom. Similarly, at least three red edges join A_2 to the bottom. Since we have now six red edges, at least two of them must terminate at the same vertex B_i of the bottom. Then $A_1A_2B_i$ is a red triangle, a contradiction. Thus all edges of both pentagons must be of the same color.

The same argument works if the top and the bottom shapes are n-gons with an odd number n. However, if n is even, the conclusion is false. For instance, a counterexample is obtained by painting the top edges red, the bottom edges green, and the edge A_iB_j red if i and j have the same parity and green otherwise.

Problem SJ1.7. Let C denote the sum of the digits of B, and let $N = 4444^{4444}$.

$$N = 4444^{4444} = 4444^{3(1481)+1} = 4444^{3(1481)} \cdot 4444,$$
$$N \equiv (4+4+4+4)^{3(1481)} \cdot (4+4+4+4)$$
$$\equiv (7^3)^{1481} \cdot 7 \equiv 1^{1481} \cdot 7 \equiv 7 \pmod 9,$$

since $7^3 = 343 \equiv 3+4+3 \equiv 1 \pmod 9$.

Thus $C \equiv B \equiv A \equiv N \equiv 7 \pmod 9$. On the other hand, $\log_{10} N = 4444 \cdot \log_{10} 4444 \approx 16210.71 < 16211$. Therefore N has at most 16211 digits, and hence $A \le 9 \cdot 16211 = 145899$. The number 99999 clearly has the largest sum of digits among all positive integers that are less than or equal to 145899, and so $B \le 5 \cdot 9 = 45$. Now, 39 has the largest sum of digits among natural numbers that do not exceed 45. Hence $C \le 3+9 = 12$. Since $C \equiv 7 \pmod 9$, and $C \le 12$, $C = 7$.

Problem SJ1.8. Let us denote each of the Blue Iris knights by a $+$, and each of the Purple Iris knights by a $-$. An arrangement of the knights around the round table can be recorded by a cycle similar to the following one:

$$---++-+++-+--+--++++ \qquad (3)$$

Now, remove all men but the last one from each group of knights of the same clan who sit together. Then we will obtain an arrangement similar to the cycle shown here:

$$-+-+-+-+-+ \qquad (4)$$

Clearly, there will be the same number of knights from both clans in this last arrangement, and hence the number of remaining knights is even, say, $2k$. On the other hand, to replace the cycle (3) by cycle (4), we removed exactly all those knights who had a friend sitting to the right of them. Hence the number of the removed knights is equal to the number of the remaining ones, and thus the total number is $2k + 2k = 4k$. Since $1386 = 4 \cdot 346 + 2$, the total number of knights around the table cannot be 1386.

Problem SJ1.9. The simplest three-term arithmetic sequence is 1, 2, 3. Its product is 6; if we multiply each term by 6, then the result will be again an arithmetic sequence and the product of its terms will be $6^4 = 36^2$.

Problem SJ1.10 We work as follows:
$$4^{545} + 545^4 = (2^{545})^2 + 2 \cdot (2^{545})(545^2) + (545^2)^2 - 2 \cdot (2^{545})(545^2)$$
$$= (2^{545} + 545^2)^2 - (2^{546})(545^2)$$
$$= (2^{545} + 545^2)^2 - ((2^{273})(545))^2$$
$$= (2^{545} + 545^2 - 545 \cdot 2^{273})(2^{545} + 545^2 + 545 \cdot 2^{273}).$$

Thus the number is composite.

Problem SJ1.11. Let us label squares from left to right with numbers 1 through 100. Let's call a square *winning* if the player who moves the marker from this square wins, and let's call a square *losing* if the player who moves the marker from this square loses. It's clear that if the only squares reachable from a given square are winning, the square is losing; but if from a given square it is possible to move the marker to a losing square, then the square is winning. Obviously, square 100 is losing; therefore, square 99 is winning, so square 98 is losing, and so on, up to square 91 which is winning. Now, all squares 81 through 90 are winning since from each one of them it is possible to move the marker to a losing square — if the marker is in square 81 it can be moved 11 squares to the losing square 92; the marker in square 82 can also be moved to the same square 92, etc. In the same way, we conclude that among squares from 71 to 80 all odd squares are winning while all even squares are losing, and all squares from 61 to 70 are winning. Continuing in the same manner, we conclude that square number 1 is winning, so the first player has a winning strategy — at each turn, he should place the marker in a losing square.

Problem SJ1.12. (a) We know that
$$1 + 2 + 3 + \cdots + 99 = \frac{99 \cdot 100}{2} = 99 \cdot 50;$$
since this number is not divisible by 4, it is impossible to construct a square using all the given segments.

(b) Let's pair up the given segments as follows: $(1, 98), (2, 97), (3, 96), \ldots,$ $(49, 50)$. The total length of each pair is 99; thus we can form 50 segments of length 99 each — 49 of them will be formed by two segments each, and the last one will be a single segment of length 99. Clearly, we can construct a rectangle of any size of the form $99n \times 99(25 - n)$, where $n = 1, 2, \ldots, 24$.

Problem SJ1.13. Let's draw four vertical and four horizontal lines which split the rectangle $ABCD$ into 25 congruent squares. Both points P and Q are vertices of two of these small squares; moreover, PQ and DQ each is a hypotenuse of a right triangle with legs 1 and 4, and one of these triangles can be obtained from the other by a 90-degree rotation about the point Q. Therefore PDQ is an isosceles right triangle with the right angle at Q.

Problem Set SJ2

Homework Problems

Problem SJ2.1. Look at the last (rightmost) digits of the numbers 2^1, 2^2, 2^3, 2^4, 2^5, 2^6, 2^7, 2^8. What is the last digit of 2^{20}, of 2^{125}, and of $2^{1234567}$?

Problem SJ2.2. Find the final *two* digits of $7^1, 7^2, 7^3, 7^4, 7^5, 7^{41}$, and 7^{12345}.

Problem SJ2.3. Suppose we calculate exactly the value of $(732.15)^{431}$. What are the last three digits in the decimal expansion, before it terminates with zeros?

Problem SJ2.4. How many ending zeros are there if you expand 1000! completely? ($1000! = 1000 \times 999 \times \cdots \times 3 \times 2 \times 1$.)

Problem SJ2.5. Provide a complete factorization of 23! into primes.

Problem SJ2.6. What is the remainder when 3^{1969} is divided by 7?

Problem SJ2.7. Find all numbers n that satisfy the following condition: among the numbers 1, 2, 3, ..., 999, 1000, there are exactly ten of them such that the sum of the number's digits is exactly n.

Problem SJ2.8. Does there exist a triangle such that the sum of two of its altitudes is larger than the sum of the two corresponding bases?

Problem SJ2.9. Four people move along a road: one is driving a car, another is riding a motorcycle, a third is riding a moped, and a fourth is riding a bicycle. Each maintains a constant speed. The car driver was next to the moped driver at 12:00 pm, was next to the bicyclist at 2:00 pm, and was next to the motorcyclist at 4:00 pm. The motorcyclist was next to the moped driver at 5:00 pm, and next to the bicyclist at 6:00 pm. At what time was the bicyclist next to the moped driver?

Problem SJ2.10. Is it possible to place numbers 1, 2, 3, 4, 5, 6, 7, and 8 at the vertices of a regular octagon in such a way that the sum of every three neighbouring numbers is larger than 11? How about larger than 13?

In-Class Problems

Problem SJ2.11. Suppose we are working in base-12, where the "digits" are 0, 1, 2, 3, 4, 5, 6, 7, 8, 9, $a = 9+1$, and $b = 9+2$. How many last zeros will there be if you expand $1a5!$ completely?

Problem SJ2.12. Show that $2222^{5555} + 5555^{2222}$ is divisible by 7.

Problem SJ2.13. Find the last digit of the number $1^2 + 2^2 + 3^2 + \cdots + 99^2$.

Problem SJ2.14. The Fibonacci sequence is defined as follows: $F_0 = 0$, $F_1 = 1$, and if $n > 1$, $F_n = F_{n-1} + F_{n-2}$. The first few terms in the sequence are: 0, 1, 1, 2, 3, 5, 8, 13, 21, 34, ... For which n is F_n divisible by 2? by 3? by 5? by 7? by 11?

Problem SJ2.15. $ABCD$ is a quadrilateral (not necessarily convex) of area 1. The midpoints of sides AB, BC, CD, and AD are denoted K, L, M, and N, respectively. Find the area of $KLMN$.

Problem SJ2.16. Alice has 12 cookies and Bob has 9 cookies. Charley, who has no cookies, pays Alice and Bob 42 cents to share their cookies. Each one of them eats one-third of the cookies.

Bob says that he and Alice should split 42 cents evenly, and Alice thinks that she should get 24 cents and Bob should get 18 cents. What is the fair division of 42 cents between Alice and Bob?

Problem SJ2.17. Can the product of two consecutive natural numbers be equal to the product of two consecutive even numbers?

Problem SJ2.18 Suppose $f(x) = x^2 + 12x + 30$. Solve the equation
$$f(f(f(f(f(x))))) = 0.$$

Solutions to Problem Set SJ2

Problem SJ2.1. It is clear that we only need to keep track of the last digit. If we look at the last digit of powers of two beginning with 2^1 we have: $2, 4, 8, 6, 2, 4, 8, 6, \ldots$. Therefore the last digit of 2^n for $n > 0$ is equal to 2, 4, 8, or 6 depending on whether $n \equiv 1, 2, 3$, or 0 modulo 4. Since $20 \equiv 0$ modulo 4, the last digit of 2^{20} is 6. Since $125 = 31 \cdot 4 + 1 \equiv 1$ modulo 4, the last digit of 2^{125} is 2. Finally, $1234567 = 308641 \cdot 4 + 3 \equiv 3$ modulo 4 and hence the last digit of $2^{1234567}$ is 8.

Problem SJ2.2. Here we need to keep track of the last two digits. The last two digits of powers of 7 beginning with 7^1 are 07, 49, 43, 01, 07, 49, 43, 01, $\ldots$. The last two digits cycle every four terms, similarly to the powers of 2 in the previous problem. $41 \equiv 1$ and $12345 \equiv 1$ modulo 4, so both 7^{41} and 7^{12345} have 07 as their last digits.

Problem SJ2.3. There is nothing special about the decimal point. If the number were 73215 raised to powers, the final three digits are the same. So we need to find the pattern of the last three digits, and it's pretty simple: 215, 225, 375, 625, 375, 625, 375, 625, $\ldots$. Beginning with the third power, the last three digits are 375 and 625, depending on whether the power is odd or even. Since 431 is odd, the last three digits are 375.

Problem SJ2.4. A zero will appear at the end every time we have a factor of $10 = 2 \times 5$ in the product. The factors of 2 will occur in the product much more often than the factors of 5, so we just need to count the factors of 5 in the numbers from 1 to 1000. There are 200 multiples of 5, 40 multiples of 25, 8 multiples of 125, and one multiple of 625. Thus there will be $200 + 40 + 8 + 1 = 249$ trailing zeros.

Problem SJ2.15. We just need to factor all the numbers from 2 to 23 and combine them. So we have:

$$\begin{aligned} 23! &= (2)(3)(2^2)(5)(2 \cdot 3)(7)(2^3)(3^2)(2 \cdot 5)(11)(2^2 \cdot 3)(13) \\ &\quad (2 \cdot 7)(3 \cdot 5)(2^4)(17)(2 \cdot 3^2)(19)(2^2 \cdot 5)(3 \cdot 7)(2 \cdot 11)(23) \\ &= (2^{19})(3^9)(5^4)(7^3)(11^2)(13)(17)(19)(23). \end{aligned}$$

Problem SJ2.6. If we look at powers of 3 modulo 7 beginning with 3^0, we see that there is a repeating cycle of length 6, namely, 1, 3, 2, 6, 4, 5, 1, 3, 2, 6, 4, 5, $\ldots$. Since $1969 \equiv 1$ modulo 6, the remainder is 3.

Problem SJ2.7. The only numbers that satisfy the conditions are $n = 3$, and $n = 24$.

Indeed, there are exactly 10 numbers with digit sum equal to 3: 3, 30, 300, 12, 21, 120, 210, 102, 201, 111.

Likewise, there are exactly 10 numbers with digit sum equal to 24: 888, 996, 969, 699, 987, 978, 879, 897, 789, 798.

To show that there are no more values n that satisfy the given condition we can check all the remaining numbers between 1 and 27, since clearly n cannot be larger than 27. To simplify this task, we notice that if a number has decimal representation xyz and all three digits are distinct, then five other three-digit numbers with the same digits have the same digit sum. Thus if a number n can be written in two different ways as $n = x + y + z$, then there are at least 12 different three-digit numbers with the digital sum n.

We have
$$5 = 0 + 1 + 4 = 0 + 2 + 3;$$
$$6 = 0 + 2 + 4 = 1 + 2 + 3;$$
$$7 = 0 + 2 + 5 = 1 + 2 + 4;$$
$$8 = 0 + 1 + 7 = 1 + 2 + 5;$$
$$9 = 0 + 1 + 8 = 1 + 2 + 6;$$
$$10 = 0 + 1 + 9 = 1 + 2 + 7;$$
$$11 = 0 + 2 + 9 = 1 + 2 + 8;$$
$$12 = 0 + 3 + 9 = 1 + 2 + 9;$$
$$13 = 0 + 4 + 9 = 1 + 3 + 9;$$
$$14 = 0 + 5 + 9 = 1 + 4 + 9;$$
$$15 = 0 + 6 + 9 = 1 + 5 + 9;$$
$$16 = 0 + 7 + 9 = 1 + 6 + 9;$$
$$17 = 0 + 8 + 9 = 1 + 7 + 9;$$
$$18 = 1 + 8 + 9 = 2 + 7 + 9;$$
$$19 = 2 + 8 + 9 = 3 + 7 + 9;$$
$$20 = 3 + 8 + 9 = 4 + 7 + 9;$$
$$21 = 4 + 8 + 9 = 5 + 7 + 9;$$
$$22 = 5 + 8 + 9 = 6 + 7 + 9.$$

Thus we need only check $n = 1, 2, 4, 23, 25, 26, 27$.

If $n = 1$, there are only 4 numbers with digit sum 1: 1, 10, 100, and 1000.

If $n = 2$, there are 6 numbers since $2 = 0 + 0 + 2$ (yielding 3 numbers), or $2 = 0 + 1 + 1$ (yielding 3 numbers again).

If $n = 4$, $4 = 0 + 0 + 4$ (3 numbers), or $4 = 0 + 1 + 3$ (6 numbers), or $4 = 0 + 2 + 2$ (3 numbers), or $4 = 1 + 1 + 2$ (3 numbers). Thus there are 15 numbers with digit sum 4.

Similarly, $23 = 6+8+9$ (6 numbers), or $23 = 7+7+9$ (3 numbers), or $23 = 7+8+8$ (3 numbers), or $23 = 5+9+9$ (3 numbers), so that there are 15 numbers with digit sum 23.

For $n = 25$, we have $25 = 7+9+9$ (3 numbers), or $25 = 8+8+9$ (3 numbers), for a total of 6 numbers.

If $n = 26$, $26 = 8+9+9$, yielding exactly 3 numbers.

Finally, $27 = 9+9+9$, so that there is exactly 1 number with this digit sum.

Problem SJ2.8. Let ABC be a triangle, let AD be the altitude perpendicular to BC, and BE the altitude perpendicular to AC, and let a, m, b, n be the lengths of BC, AD, AC, BE, respectively.

The area of triangle $ABC = am/2 = bn/2$, so $am = bn$, and thus $a/b = n/m = k$, whence $a = kb$ and $n = km$.

Suppose that $(m+n) > (a+b)$. Then $(m+km) > (kb+b)$, so that $m(1+k) > b(1+k)$, and hence $m > b$, i.e. the length of AD is larger than the length of AC, but this is impossible since either $AC = AD$ (if the angle ACB is a right angle), or else ADC is a right triangle with hypotenuse AC in which the length of AD is less than the length of AC.

Problem SJ2.9. Consider a coordinate system with the horizontal axis representing time (in hours, with 0 corresponding to 12:00 pm) and the vertical axis representing distance.

Since all people move with constant speed, their movements will be represented by straight lines. Let lines k, l, m, n represent the car, moped, bicycle, and motorcycle, respectively. Then

k passes through points $(0,a)$, $(2,b)$, and $(4,c)$;

l passes through points $(0,a)$ and $(5,d)$;

m passes through points $(2,b)$ and $(6,e)$;

n passes through points $(4,c)$, $(5,d)$, and $(6,e)$.

We need to find the x-coordinate of the point of intersection of the lines l and m.

Now, the slope of k is $(b-a)/2 = (c-b)/2$, so $2b = a+c$.

The slope of n is $(d-c)/1 = (e-d)/1$, so $2d = c+e$.

The slope of l is $(d-a)/5$, and its equation is $y = ((d-a)/5)x + a$.

The slope of m is $(e-b)/4$, and its equation is $y-b = ((e-b)/4)(x-2)$.

Thus l and m intersect at a point where

$$((d-a)/5)x + a = ((e-b)/4)(x-2) + b,$$

whence

$$x = (10(3b - e - 2a))/(4d - 4a - 5e + 5b)$$
$$= (10(6b - 2e - 4a))/(8d - 8a - 10e + 10b)$$
$$= (10(3a + 3c - 2e - 4a))/(4c + 4e - 8a - 10e + 5a + 5c)$$
$$= (10(-a + 3c - 2e))/(3(-a + 3c - 2e)) = 10/3.$$

Hence the bicyclist and the moped met at 3:20 pm.

Problem SJ2.10. (a) It is possible: Place numbers at the vertices in the following order: 1, 5, 7, 2, 4, 6, 3, 8.

(b) We'll prove that this is impossible by contradiction. Suppose that the arrangement a, b, c, d, e, f, g, h satisfies the given condition, so that we have the following eight inequalities:

$$a + b + c \geq 14; \quad b + c + d \geq 14;$$
$$c + d + e \geq 14; \quad d + e + f \geq 14;$$
$$e + f + g \geq 14; \quad f + g + h \geq 14;$$
$$g + h + a \geq 14; \quad h + a + b \geq 14.$$

Adding all eight inequalities we get $3(a + b + c + d + e + f + g + h) = 3(1 + 2 + 3 + 4 + 5 + 6 + 7 + 8) = 108 \geq 8 \cdot 14 = 112$, a contradiction.

Problem SJ2.11. In base-12 arithmetic, we will have a terminal zero every time the number contains a factor of $12 = 2^2 \cdot 3$. Thus we need to count the number of factors of 2 and 3 in $1a5!$. In base-10, the number $1a5$ is $1 \cdot 12^2 + 10 \cdot 12 + 5 = 269$. Similarly to Solution SJ2.4 above, we find that there are $134 + 67 + 33 + 16 + 8 + 4 + 2 + 1 = 265$ factors of 2 and therefore 132 factors of 4; also, there are $89 + 29 + 9 + 3 + 1 = 131$ factor of 3. Thus there will be 131 terminal zeroes when $1a5!$ is written in base-12.

Problem SJ2.12. All we need to do is calculate everything modulo 7. We can easily see that $2222 \equiv 3$ and $5555 \equiv 4$ modulo 7. In Problem SJ2.6 we have already looked at powers of 3 modulo 7. Now $5555 \equiv 5$ modulo 6, so $2222^{5555} \equiv 3^{5555} \equiv 5$ modulo 7.

Let us now look at powers of 4 modulo 7. Beginning with 4^0, we have a cycle of length 3, namely 1, 4, 2, 1, 4, 2, ... Since $2222 \equiv 2$ modulo 3, we get $5555^{2222} \equiv 4^{2222} \equiv 2$ modulo 7.

Adding 5 and 2 gives a number that is 0 modulo 7, so $2222^{5555} + 5555^{2222}$ is divisible by 7.

Problem SJ2.13. We need to keep track of the last digits only. For the numbers from 1 to 10, the last digits of their squares are 1, 4, 9, 6, 5, 6, 9, 4, 1, 0. If we add them together, the final digit is 5. Similarly, the last digit of $11^2 + 12^2 + \cdots + 20^2$ will be 5, and so on. We can express it modulo 10 as follows:

$$\begin{aligned}
1^2 + 2^2 + \cdots + 99^2 &= (1^2 + 2^2 + \cdots + 10^2) \\
&+ (11^2 + 12^2 + \cdots + 20^2) \\
&+ (21^2 + 22^2 + \cdots + 30^2) \\
&+ \cdots \\
&+ (91^2 + 92^2 + \cdots + 99^2) \\
&\equiv 5 + 5 + 5 + \cdots + 5 + (1 + 4 + 9 + 6 + 5 + 6 + 9 + 4 + 1) \\
&\equiv 5 \cdot 10 \equiv 0.
\end{aligned}$$

Thus the last digit is 0.

Problem SJ2.14. It's easy to find the patterns. Just write down the Fibonacci numbers modulo 2, 3, 5, 7, and 11 until they repeat:

Modulo 2: 0, 1, 1, 0, 1, 1, ...;
Modulo 3: 0, 1, 1, 2, 0, 2, 2, 1, 0, 1, 1, ...;
Modulo 5: 0, 1, 1, 2, 3, 0, 3, 3, 1, 4, 0, 4, 4, 3, 2, 0, 2, 2, 4, 1, 0, 1, 1, ...;
Modulo 7: 0, 1, 1, 2, 3, 5, 1, 6, 0, 6, 6, 5, 4, 2, 6, 1, 0, 1, 1, ...;
Modulo 11: 0, 1, 1, 2, 3, 5, 8, 2, 10, 1, 0, 1, 1, ...

From the sequences above, we see that F_n is divisible by 2 if n is divisible by 3; F_n is divisible by 3 if n is divisible by 4; F_n is divisible by 5 if n is divisible by 5; F_n is divisible by 7 if n is divisible by 8; and F_n is divisible by 11 if n is divisible by 10.

Problem SJ2.15. Triangle AKN is similar to the triangle ABD with the coefficient of proportionality of $\frac{1}{2}$. Thus its area is $\frac{1}{4}$ of the area of ABD. Likewise, the area of CML is $\frac{1}{4}$ of the area of CDB. Therefore the sum of the areas of AKN and CLM is $\frac{1}{4}$ of the area of $ABCD$. Similarly, the sum of the areas of BLK and DNM is $\frac{1}{4}$ of the area of $ABCD$. Hence the area of $KLMN$ is $\frac{1}{2}$.

Problem SJ2.16. Since Charley ate seven cookies and paid 42 cents, each cookie costs 6 cents. Thus Alice, who sold five cookies, should get 30 cents, and Bob should get 12 cents.

Problem SJ2.17. Suppose that $n > 0$ and $n(n+1) = 2m(2m+2)$. If $n < 2m$, then $(n+1) < (2m+1) < (2m+2)$, and so $n(n+1) < 2m(2m+2)$, a contradiction. If $n = 2m$, then $(n+1) = (2m+1) < (2m+2)$, so $n(n+1) < 2m(2m+2)$, a contradiction. If $n > 2m$, then $n \geq (2m+1)$ (since n is an integer), and therefore $(n+1) \geq (2m+2)$. Hence again, $n(n+1) > 2m(2m+2)$, a contradiction.

Hence the product of two consecutive natural numbers cannot be equal to the product of two consecutive even integers.

Problem SJ2.18. We can write $f(x) = x^2 + 12x + 30 = x^2 + 12x + 36 - 6 = (x+6)^2 - 6$, and so $f(f(x)) = f((x+6)^2 - 6) = (((x+6)^2 - 6) + 6)^2 - 6 = (x+6)^4 - 6$.

Similarly, $f(f(f(x))) = (x+6)^8 - 6$, and so on.

Thus $f(f(f(f(f(x))))) = (x+6)^{32} - 6 = 0$, whence $(x+6) = 6^{1/32}$, and so $x = 6^{1/32} - 6$.

Problem Set SJ3

Homework Problems

Problem SJ3.1. Find a number that uses each of the digits 1 through 9 exactly once, such that the number formed by looking at the leftmost n digits is divisible by n, $n = 2, 3, \ldots, 9$.

Problem SJ3.2. Prove that if p and $p^2 + 2$ are prime, then so is $p^3 + 2$.

Problem SJ3.3. A laser beam is shot from the southwest corner of a square made out of a reflective material. The beam is aimed at the northeast corner; if it exactly hits this corner or any other corner, it is absorbed. If it hits any other place, it bounces. It turns out that the aim was not quite perfect, and the beam bounced 2009 times before it got absorbed.

(a) Assuming that the aim was as good as possible (i.e., as close to 45 degrees as can be), what was the angle?
(b) How many angles would yield 2009 bounces?
(c) What if it was a cube instead of a square?

Problem SJ3.4. (a) Is there a set of points in the plane that meets every circle in exactly two points?
(b) Is there a set of points in the plane that meets every circle of radius 1 in exactly two points?

Problem SJ3.5. Which of the two numbers is bigger: $9^{9^{9^{\cdot^{\cdot^{\cdot}}}}}$ (one hundred exponents) or $9!!!\ldots!$ (one hundred factorials)?

In-Class Problems

Problem SJ3.6. If the square of a positive integer ends with 69, prove that either the previous or the next perfect square ends with 96.

Problem SJ3.7. A battalion 20 miles long advances 20 miles. During this time, a messenger on a horse travels at a constant speed from the rear of the battalion to the front, and immediately turns around and ends up precisely at the rear of the battalion as it completes the 20-mile journey. How far has the messenger traveled?

Problem SJ3.8. A three-digit number has the sum of its digits equal to 7. Show that if the number is divisible by 7, then the tens digit is equal to the units digit.

Problem SJ3.9. Consider all nine-digit integers (written in decimal) with distinct digits and no 0 digit. Find the GCD of all of them.

Problem SJ3.10. Show that in any convex polygon with 21 sides, there are two diagonals that are either parallel or form an angle that is less than $1°$.

Problem SJ3.11. Let $\overleftrightarrow{AB}$ and $\overleftrightarrow{CD}$ be two lines intersecting at point O. Let $\overrightarrow{OP}$ be the bisector of $\angle AOC$, $\overrightarrow{OT}$ the bisector of $\angle POB$, and $\overrightarrow{OR}$ the bisector of $\angle TOD$. If $\angle POR = 25°$, find $\angle AOC$ and $\angle AOD$.

Solutions to Problem Set SJ3

Problem SJ3.1. The answer is unique; it is 381,654,729.

The solution requires some case testing, but if you are careful it is not too difficult.

Let d_k denote the k-th digit of this number, and let $[i \ldots j]$ denote the base-10 number formed by digits d_i through d_j. For example, for the answer above, $d_6 = 4$ and $[2 \ldots 5] = 8165$.

Observe that the number we seek has the following properties:

(a) d_2, d_4, d_6, d_8 are even.
(b) $d_5 = 5$.
(c) The sums $d_1 + d_2 + d_3$, $d_4 + d_5 + d_6$, and $d_7 + d_8 + d_9$ are multiples of 3.
(d) $[3 \ldots 4]$ is a multiple of 4.
(e) $[6 \ldots 8]$ is a multiple of 8.

Looking at $[3 \ldots 4]$, we observe that d_4 must be either 2 or 6—since d_3 is odd, if d_4 is 4 $[3 \ldots 4]$ wouldn't be a multiple of 4. (For example, 78 is not a multiple of 4 but 72 is.)

By the same reasoning, d_8 must be either 2 or 6 (since if $[6 \ldots 8]$ is a multiple of 8, then surely $[7 \ldots 8]$ is a multiple of 4). Thus d_4, d_8 are 2 or 6; and d_2, d_8 are 4 or 8.

This leads to just four possible arrangements of even digits: $x4x258x6x$, $x4x654x2x$, $x8x254x6x$, and $x8x654x2x$.

For each of these, there is a fairly limited number of choices for d_1, d_3, d_7 dictated by properties (c) and (e) above. The final test used is the fact that $[1 \ldots 7]$ must be divisible by 7. (We don't have to worry about d_9—why?).

Problem SJ3.2. If $p = 3$, then indeed $p^2 + 2 = 11$ and $p^3 + 2 = 29$ are primes. However if $p > 3$, the remainder upon dividing p by 3 is 1 or 2; this means that the remainder on dividing p^2 by 3 must be 1, and it implies that $p^2 + 2$ is divisible by and greater than 3, so it cannot be prime. Thus the only prime that satisfies the conditions is 3.

Problem SJ3.3. (a) Assume the square is a unit square and let the slope of the laser beam be a/b, where a and b are relatively prime positive integers. Now imagine a line with this slope extending northeast from the origin $(0,0)$. The first lattice point that this line hits after leaving the origin is (b, a). Each horizontal or vertical grid line that this line hits corresponds to a bounce;

we just need to count them. It is easy to see that the line will intersect $(a-1)$ vertical lines and $(b-1)$ horizontal lines, so the number of bounces is $(a+b-2)$. Our problem is now reduced to finding the slope a/b such that a and b are as close in value as possible, are relatively prime, and add up to 2011. Since consecutive integers are relatively prime, the two solutions are $1006/1005$ or $1005/1006$; the answers are the inverse tangents of these fractions.

(b) We need to find all relatively prime positive integers a and b such that $a+b = 2011$. Notice that for any two integers u, v, if an integer d divides both u and v then surely d divides both u and $u+v$, and conversely. Consequently, if u is relatively prime to v, then u is relatively prime to $u+v$.

Applying this to the problem at hand, if a is relatively prime to b and $a+b = 2011$, then a must be relatively prime to 2011, and conversely. So we need only count the number of integers relatively prime to 2011. Since 2011 is prime, our answer is 2010.

All the slopes $1/2010, 2/2009, 3/2008, 4/2007, \ldots, 2010$ yield appropriate angles.

(c) Here is a partial solution, or rather some hints. We use the same idea as above, but in three dimensions. Our line goes from the origin $(0,0,0)$ to (a,b,c), where a, b, c have no common factors (for example, $a=5$, $b=10$, $c=3$ is appropriate: even though a and b share a common factor 5, there are no common factors for all three numbers). Now we count $a+b+c-3$ bounces. Hence $a+b+c = 2012$. However, there are a few tricky things that need to be thought about, such as counting the solutions, dealing with one of the three values equalling zero (not illegal), etc.

Problem SJ3.4. (a) The answer is no. The purported set S meets a certain circle O in two points, say A and B. For every point X in the plane which doesn't lie on the line AB, there's a circle through A, B, and X, and our set already meets that circle at A and B; thus it cannot contain any such point X. So S is confined to the line AB, which is absurd since there are many circles that don't meet that line.

(b) Here the answer is yes. Consider a set consisting of infinitely many horizontal lines spaced two units apart. Each circle of radius 1 either meets such a line in exactly two points or else it is tangent to two of these lines; in this case, again, it meets our set in two points.

Problem SJ3.5. Let $a_0 = b_0 = 9$ and $a_{n+1} = 9^{a_n}$, $b_{n+1} = b_n!$. The problem is to compare a_{100} and b_{100}.

Let's prove by induction that for $n > 0$ we have $a_n > 2b_n \log_9 b_n$, where the $\log_9$ is to the base 9. This is a nice example of a familiar situation: strengthening a statement sometimes makes it easier to prove by induction — you have to prove more, but you also have more to work with.

So $a_1 = 9^9$, which is more than $8^9 = 2^{27} = (2^7)(2^{10})^2$, and that's more than $128{,}000{,}000$ (since $2^{10} > 1{,}000$ should be a familiar fact). On the other

Set SJ3

hand, $9! = 362,880$ and $2(9! \cdot \log_9 9!)$ is certainly less than $2(9! \times 9)$ (since $9! < 9^9$), and therefore less than 8 million. Anyway it should be clear that $a_1 > 2b_1 \log_9 b_1$.

Now suppose that $n > 0$ and $a_n > 2b_n \log_9 b_n$. Then

$$a_{n+1} = 9^{a_n} > 9^{2b_n \log_9 b_n} = (b_n)^{2b_n} = ((b_n)^{b_n})^2 > (b_n!)^2.$$

We need to show that this is more than $2b_{n+1} \log_9 b_{n+1} = 2(b_n!) \log_9 (b_n!)$. Thus it suffices to show that $b_n! > 2 \log_9(b_n!)$ which is certainly true ($x > 2 \log_9 x$ is equivalent to $9^x > x^2$, which is true as soon as x is larger than 2 — this can easily be proved by induction).

Problem SJ3.6. Let $[ab \ldots c]$ denote the number $a \times 10^n + b \times 10^{n-1} + \cdots + c$ (for example, $[203] = 2 \times 10^2 + 0 \times 10^1 + 3$).

Let $n = [x_1 x_2 \ldots x_k ab]$. Clearly, the last two digits of n^2 are the same as the last two digits of $[ab]^2$. So we have $[ab]^2 = [y_1 y_2 \ldots y_l 69]$. Thus $b = 3$ or $b = 7$ and hence

$$ab \in \{13, 17, 23, 27, 33, 37, 43, 47, 53, 57, 63, 67, 73, 77, 83, 87, 93, 97\}.$$

But since $[ab]^2$ ends in 69, it follows easily that $ab \in \{13, 37, 63, 87\}$. Since $14^2 = 196$, $36^2 = 1296$, $64^2 = 4096$, and $86^2 = 7396$, the statement is now proved.

Problem SJ3.7. Let d be the distance travelled by the battalion until the messenger gets to the front of the battalion, and let t be the amount of time it takes the messenger to get to the front of the battalion. Let x be the rate (in mph) at which the messenger is travelling, and let y be the rate (in mph) at which the battalion is moving. Then $t = \frac{d}{y}$ and also $t = \frac{20+d}{x}$, so $\frac{d}{y} = \frac{20+d}{x}$, or equivalently $\frac{y}{x} = \frac{d}{20+d}$.

Now let T be the time it takes the messenger to get back to the rear of the battalion. Then $T = \frac{20-d}{y} = \frac{d}{x}$, so $\frac{y}{x} = \frac{20-d}{d}$. Therefore $\frac{d}{20+d} = \frac{20-d}{d}$, so $d^2 = 400 - d^2$, and thus $d = \sqrt{200} = 10\sqrt{2}$.

Hence, the messenger has travelled a distance of $20 + 2d = 20 + 20\sqrt{2}$ miles.

Problem SJ3.8. Let $N = [xyz]$ be the three-digit number with $x \neq 0$. By hypothesis, $x + y + z = 7$, and 7 divides $100x + 10y + z$. From the first equation, $y + z = 7 - x$ and since $x \geq 1$, we have that $y + z \leq 6$.

Since 7 divides $98x + 7y$, it must also divide $100x + 10y + z - (98x + 7y) = 2x + 3y + z = (x + y + z) + x + 2y = 7 + x + 2y$. Hence 7 divides $x + 2y$. But $x + 2y = 7 - y - z + 2y = 7 + y - z$, and so 7 divides $y - z$. Since y and z are decimal digits, $-9 \leq y - z \leq 9$, so it must be that $y - z = -7$, 0, or 7. If $y - z = -7$, then $y + z = 7 + 2y \geq 7$. If $y - z = 7$, then again, $y + z = 7 + 2z \geq 7$. But we know that $y + z \leq 6$.

This leaves us with the only possibility, $y = z$.

Problem SJ3.9. Among those numbers we find 987654321 and 987654312, which differ by 9, so the GCD divides 9. On the other hand, 9 certainly does divide all of the numbers by the well-known criterion for divisibility by 9. So the answer is 9.

Problem SJ3.10. Let P be a point in the same plane as our 21-gon. Through the point P, draw lines parallel to all $\frac{21 \cdot 18}{2} = 189$ diagonals of the polygon. These lines form $2 \cdot 189 = 378$ angles. Since the sum of all these angles is 360 degrees, we deduce that at least one of these angles must be less than 1 degree.

Problem SJ3.11. The main difficulty here is to determine the relative positions of the rays and to come up with a helpful picture. Let's consider two possibilities: $\angle AOC \leq 90°$, and $\angle AOC > 90°$.

Assume first that $\angle AOC \leq 90°$, and let $\angle AOC = 2x$. Then $x \leq 45°$. Since $\overrightarrow{OP}$ bisects $\angle AOC$, so $\angle AOP = \angle POC = x$, and so $\angle POB = 180 - x$. Since $\overrightarrow{OT}$ bisects $\angle POB$, so $\angle POT = \angle TOB = \frac{180-x}{2}$. Note that $\angle POT - \angle POC = \frac{180-x}{2} - x = \frac{180-3x}{2} > 0$ since $x \leq 45°$, and hence ray $\overrightarrow{OC}$ is between rays $\overrightarrow{OP}$ and $\overrightarrow{OT}$, and $\angle COT = \frac{180-3x}{2}$. Then $\angle TOD = 180 - \angle COT = 180 - \frac{180-3x}{2} = \frac{180+3x}{2}$, and since $\overrightarrow{OR}$ bisects $\angle TOD$, $\angle TOR = \frac{180+3x}{4}$. Then we must have $25 = \angle POR = x + \frac{180-3x}{2} + \frac{180+3x}{4} = \frac{540+x}{4} > 135$, a contradiction. This situation is impossible.

Now assume instead that $\angle AOC$ is more than $90°$. We can illustrate the situation as in the diagram on the right. Let $\angle AOD = \angle COB = x$, $\angle AOR = y$, $\angle POT = z$, $\angle TOC = w$, as on the diagram. We then have

$y + 25 = z + w$ (since ray OP bisects $\angle AOC$), (5)

$z = w + x$ (since ray OT bisects $\angle POB$), (6)

$x + y = 25 + z$ (since ray OR bisects $\angle TOD$), (7)

$x + y + 25 + z + w = 180$ (since $\angle DOC = 180°$). (8)

From (6) we can replace z in (7) by $w + x$, thus obtaining $x + y = 25 + w + x$, and eliminating x, we get $y = 25 + w$. Now we can replace both z and y in (5) and (8), and we get two equations with two unknowns:

$$x + w = 50, \quad 2x + 3w = 130.$$

Solving this system we get $w = 30$ and $x = 20$. Hence $\angle AOD = 20°$, and $\angle AOC = 180° - 20° = 160°$.